THE ROCK GARDENER'S HANDBOOK

ALAN TITCHMARSH

The Rock Gardener's Handbook

CROOM HELM London & Canberra

Croom Helm Ltd, Provident House, Burrell Row,
Beckenham, Kent BR3 1AT

Croom Helm Australia Pty Ltd, First Floor,
139 King Street, Sydney, NSW 2001, Australia

Croom Helm 51 Washington Street,
Dover, New Hampshire 03820, USA

British Library Cataloguing in Publication Data

Titchmarsh, Alan
The rock-gardener's handbook.
1. Rock gardens
635.9'672 SB459

ISBN 0-7099-4302-4 PBK

Reprinted in paperback in 1985

Printed and bound in Great Britain by
Biddles Ltd, Guildford and King's Lynn

Contents

contents

List of Figures

Foreword

Modern gardens are on a much smaller scale than in former days when it was not so essential in builders' eyes to extract the maximum unit ratio per acre of valuable land. Thus small gardens need plants in proportion and alpines fit the role perfectly. Many books have been written on alpines over the years and so you may ask, why another one? Each one is different and contributes additionally to the vast subject of alpines – during the past twenty years I have been asked time and time again, 'which book can I buy to tell me all I want to know about alpines?' I confess to chuckling slightly as I look at over 100 books and floras on my shelves and reply accordingly.

You therefore need an introduction to this fascinating subject and this book by Alan Titchmarsh is a helpful beginning to further study, for once the bug has got you, you rarely recover from the disease of growing alpines, ever pressing forward to meet the many challenges of cultivation of the rarer plants. Neither do you need a rock garden in which to grow them, for there are many alternative methods, indeed, some are mentioned in this book. But do be warned: do not expect to be able to expend a minimum amount of time and effort in the creating of a rock garden and then think you can sit back to enjoy it forever. This is an ongoing hobby and whatever may have been said elsewhere, no way can rock gardening be said to be a labour-saving form of gardening. Once you start you are in it for good, unless you allow weeds to overtake and demolish whatever effort you have put into it.

If all this sounds a little dampening, it is again based on the experience of a number of years when, perhaps a magnificent display at the Chelsea Flower Show inspires the visitor to return home to create a rock garden, only to find that it takes time, thought, effort and money. Be inspired by all means, but be aware of what you are letting yourself in for!

The attraction of alpines (or rock plants – there is no difference) does not stop at seeing them in the garden – the call of their native habitats draws many people to the mountains to enjoy them in the wild, where they should be left, incidentally, as collecting plants from the wild on an indiscriminate basis is no longer an acceptable pastime in the conservation-minded age in which we live.

Foreword

I have enjoyed reading this book. I only wish that a similar publication had been available to me in my youth when I was first fired with the delights of these miniature plants that we cultivate. Alan Titchmarsh has chosen a good selection of plants that should attract and encourage the reader and this Handbook will be a useful, but certainly not repetitive, addition to the literature on the subject. I wish you joy in this absorbing hobby.

Michael Upward
Secretary
Alpine Garden Society

For Camilla
'The rainbow comes and goes
And lovely is the rose.

William Wordsworth

Introduction

They say that in the introduction to a book you should on no account apologise for its existence. So I shan't. But I will say that this book sets out to achieve a different aim to most other alpine books. Today's gardener can find a good selection of skilfully written volumes which describe in detail the thousands of different alpines that can be grown, and a good few that can't. Some of the plants are seldom seen in cultivation, and you're only likely to obtain them if you go out into the wild with a trowel and an import permit. Often the growing instructions that accompany the glowing descriptions are minimal.

In this book I've concentrated on the growing of the plants, and have recommended just a few species in each genus that I consider to be among the best. They're all plants that I've grown or seen growing and know to be available – the list of suppliers at the back of the book should help you to track down the rarities.

Writing a book about your favourite group of plants is a daunting task. All the time you are conscious that the plants must be shown in their best light – not as tricky weeds of sour constitution that demand a lot of coddling for little return. Good cultivation and plant selection should do away with such problems. Then there are the ghosts of earlier alpine writers (and a few present-day ones too) who peer critically over your shoulder – you can almost hear their sharp intakes of breath as you recommend some cultivation technique they would be sure to frown upon. But then this is one man's view of alpines, and not always one to which other growers will subscribe. The ways I've recommended are not the only ways – would that some alpine growers were more flexible in their recommendations, and more willing to believe that there are more ways than one of growing a gentian.

Having said that, I have to admit that I've learned a lot from gardeners who grew alpines long before I gurgled in my cradle, and I continue shamelessly to borrow the techniques of friends in the alpine world who are generous enough to share their secrets.

If I have any excuses to make it's that I've had little room to speak much about bulbs, shrubs and conifers, but I hope that the brief chapters towards the end of the book will at least give you some idea of the kinds that are suitable for rock gardens, banks, beds, sinks, troughs and other 'alpine gardens'.

This is essentially a factual book, but I hope it's readable too. You'll forgive me for slipping in the occasional anecdote or quote from another book, but reading about alpines is great fun, if you select the right authors, and such quotes will, I hope, be regarded as illuminating highlights rather than dull plagiarism.

Above all, the book aims to help you enjoy growing a most fascinating group of plants. They'll thrive in most gardens, even on patios or doorsteps, and will take you into a magical miniature world. When you tire of the easy ones, there are tricky blighters that will tax you for most of the year and delight you for a part of it when they burst into flower. There are a few of these in the book, but the majority need much less fuss and bother, provided you put them in the right place and the right soil at the right time. With the information I've given, I hope you'll be able to do just that.

A.T.
Beech, 1982

CHAPTER 1

Growing the Plants

Choosing the Right Spot

The traditional 'natural' rock garden was at the height of its popularity in the late nineteenth and early twentieth centuries and looked as imposing as any other piece of Victoriana. It was usually built into a slope where vast boulders were made to protrude from the earth in the fashion of a gargantuan rocky outcrop. The rocks provided plenty of crevices in which plants could grow and, coupled with a gritty soil, made for the good drainage essential to the alpines' survival.

Today few of us have gardens which are large enough to accommodate such massive geological features, or pockets large enough to pay for the rock needed to make them. But even if your flat plot and modest bank account mean that you have to curtail your ambitious landscaping ideas, there is no reason why you should not grow your alpines equally well in a situation that looks nothing like their original home, provided that it is sympathetic to their form and supplies the environmental requirements they love.

Up in the mountains where alpines grow, the atmosphere is clean and pure and the light intensity high, so it follows that in the garden most rock plants will do best in a brightly lit spot. Shade-loving natives of wooded slopes and north-facing rock crevices are exceptions to this rule and will prefer a more dappled situation, but the majority of rock plants enjoy any site which is open to the sky but sheltered from winds by a *distant* fence or trees and shrubs.

Because many of them are so obliging, rock plants, like rhubarb, are often pushed into a corner of the garden where nothing else will survive. If the site is dark and wet, or the soil as dry and unyielding as concrete, the chances are that your alpines will die. Rock gardens made directly under trees are seldom successful. For a start the trees' roots rob the soil of nutrients and water and make cultivations difficult. Added to this the plants become spindly due to lack of light, and many of them will rot off in winter when water and leaves fall from the branches. Woodland plants are at home under trees; true alpines are not.

If your garden does slope, then make use of its contours. There are alpine plants which will tolerate all aspects, and those which prefer just one: ramondas and haberleas thrive best in vertical, north-facing crevices, and kabschia saxifrages like some protection from scorching sun – try them on the east or west side. If you just want to fill your rock garden with mossy saxifrages, sempervivums, rock roses, stonecrops and yellow alyssum, make it as open to the sky as you can and they'll grow lustily.

Many gardeners are frightened of north-facing slopes. They needn't be. It's true that sheer north faces are suitable only for shade lovers, but a gentle north slope can be moulded to provide sunny patches as well as dingy grottoes. It's more difficult to make shady nooks in south-facing rock gardens where the sun reaches all plants. The rock garden at the Royal Horticultural Society's Garden at Wisley in Surrey faces north-west and provides a multiplicity of different conditions for different plants. The rock garden at Kew is on a flat site, contoured to provide a variety of habitats. Both are certainly larger than any back garden 'rockery', but to anyone about to embark on making a rock feature at home they are still a useful source of inspiration.

When you've considered the plants' requirements you'd better consider your own. A rock garden is not always easy to fit comfortably into a suburban plot or even a large country garden, and for this reason it is often thought of as rather inferior. You'll not find a rock garden in many of Britain's most famous stately gardens, and that, I suspect, is because for half a century or so rock gardens have been thought unfashionable. Sadly, they've been given a bad name by the hundreds of 'rockeries' that pop up like miniature Mount Etnas in tiny gardens all over the country.

I cannot do better than to quote the oft-repeated words of Reginald Farrer, the 'father' of English rock gardening, who in *My Rock Garden*, first published in 1907, warned against constructing 'a Drunkard's Dream of noxious cement blocks'; who exhorted gardeners not to build 'a sham grotto of clinker bricks on the lawn, and rig up nasturtiums over it to hide the places where your alpines have unanimously died', and who warned against our familiar friend the

> Almond-pudding scheme. . . You take a round bed; you pile it up with soil; you then choose out the spikiest pinnacles of limestone you can find, and you insert them thickly with their points in the air, until the general effect is that of a tipsy cake stuck with almonds. In this vast petrified porcupine nothing will grow except Welsh Poppy, Ferns, and some of the uglier Sedums.

There were two other common catastrophes: 'the Dog's Grave' and the 'Devil's Lapful'. In the dog's grave the mound is furnished with stones laid flat. 'Plants will grow on this, but its scheme is so stodgy and abhorrent to Nature that it should be discarded'. The devil's lapful is more ambitious:

> You take a hundred or a thousand cartloads of bald, square-faced boulders. You next drop them all about absolutely anyhow and you then plant things amongst them. The chaotic hideousness of the result is something to be remembered with shudders ever after.

Beware, then, of these monstrosities, and build a garden for your alpines that is easy on the eye and easy on their roots. Use materials that look at home in your garden and next to your house, and construct a feature that is in keeping with the scale and shape of its surroundings. Some experts will extol you to make a mound-shaped rock feature if your garden is flat – I will not. I would far rather see a flat bed of alpines on a light, well-drained soil, or a level raised bed where drainage is poor, than I would an incongruous cairn of rocky lumps that supposedly simulated nature. But more on construction later; first let's consider the type of soil that the plants enjoy best, and the kinds of rock that they will enjoy growing over.

Tackling the Soil

In their natural home most alpines grow in a medium that is composed mainly of rock fragments plus smaller quantities of organic matter and soil. It is ostensibly a poor mixture in which to grow, but it has certain characteristics which are vital to the alpine plant's survival. First, it is well drained; it allows excess water to escape rapidly when the spring thaw comes, but also holds on to a fair amount of moisture in summer. The moisture-retentive organic content and a surface covering of broken rocks prevent the sun from baking plant roots.

Through the winter the alpines are covered with snow. It's not the soggy coating you might expect, but a thick, dry blanket that protects the plants from extreme temperatures, and also prevents rotting due to excessive damp. When the snow melts in spring the plants get a good watering and higher temperatures to speed them on their way.

The message is clear: give most alpines a well-drained soil and a spot that is not likely to be saturated in winter. The first proviso is easy (if laborious) to achieve – heavy soil can be improved by adding copious quantities of washed grit or coarse sand, plus coarse sphagnum peat. All three will aid drainage while allowing reasonable water retention in summer.

If the situation is really severe, the soil is sticky and unworkable, and you are determined to grow good alpines, it will be worth removing the top 12in (30cm) of soil. When it's been excavated put down a 6-in (15-cm) layer of broken bricks or rubble, and then replace it with a fresh load of topsoil mixed with grit and peat.

Light, sandy soils that are well drained but prone to dry out in summer should also have moist peat forked into the top 12in (30cm). This will help them to hang on to more moisture at a time when the plants need it.

Some alpines insist on an exceptionally well-drained soil that is composed almost entirely of rock fragments. These are the plants that grow in the scree at the foot of a mountain slope. Scree beds are simple to construct in the garden, and although you will have to expend effort and funds they will greatly widen the range of plants you can grow (see page 23 for scree-bed construction).

The acidity of your soil is another factor which will govern your choice of plants. In chalky soils lime-haters such as rhododendrons, callunas, some ericas, shortias, lithospermums (lithodoras) and most members of the family Ericaceae will not do well. They will turn yellow, become stunted and eventually die. But many alpines can tolerate lime in the soil with no ill effects, so concentrate on growing these in your rock garden and make a small peat bed (see page 25) in a raised part of the garden (so that chalky drainage water does not flow into it) if you really want to try plants that abhor your conditions. Plants that will not tolerate lime are indicated as such in the A-Z section.

Gardeners on acid soil will have few problems, for the small number of plants that demand lime can be given a dressing of ground limestone or limestone chippings at planting time and subsequently each spring. Scatter one or two handfuls of lime or chippings around the plant, and work the stuff into the soil with your fingers or a hand fork.

Protecting alpines from winter damp (without a dry covering of snow) is not so easy. Many of them can survive British winters with equanimity, but some (especially those which are felted with fine hairs) involuntarily hold on to moisture, which causes their leaves and stems to rot. All alpines susceptible to winter wet can either be covered with cloches or single panes of glass, if they are grown in the rock garden, or else cultivated in pots which can be overwintered in a cold greenhouse or frame.

Excessive moisture at the roots can also be a problem in winter, so do make sure that drainage is good, even if improving matters involves a fair amount of labour. When you've finished with the soil it should feel rough and gritty in your hand. If it remains smooth and slippery when wet, or dusty when dry, the choosiest alpines won't like it one bit.

There is much controversy over the wisdom of feeding alpines.

Some gardeners claim that the addition of manure to the rock-garden soil produces more shapely and floriferous plants. Others (and here I include myself) suggest that such overfeeding produces sappy, untypical plants that may be short lived. A third school will allow you to add a little dried cow manure to the soil – nothing else. In the absence of cows in your vicinity stick to peat or well-rotted leafmould as your soil enrichment. In my experience the plants grow perfectly well without helpings of more nutritious animal waste. After all, they don't get it at home, unless you count the rare visits of a browsing chamois or a mountaineering cow.

It goes without saying that your soil should be weed free before you plant. During cultivations all thick-rooted perennial weeds should be ousted completely – any broken-off sections of root that are left behind will most certainly grow into replicas of their parents. For this reason it is worth preparing the ground a month or two in advance of planting up so that any weeds that subsequently emerge can be dealt with. Use mechanical means if you have the patience – your hands or a trowel (and a Dutch hoe for the annual weeds which will die when severed from their roots). Use chemical means if you lack the time, energy and inclination to use your hands. Glyphosate (Tumbleweed) applied *strictly* in accordance with the manufacturer's instructions will kill off most perennial weeds, though a second or third application may be necessary for those that are the bane of the rock gardener's life – oxalis, convolvulus and ground elder. I think the worst weed to have in a rock garden is horsetail or marestail (*Equisetum arvense*). Its roots seemingly push down for miles and are impossible to eradicate once they have a foothold. Don't make your rock garden on ground that already plays host to this decorative but pernicious weed.

Finally, when you are improving and cultivating the soil, don't stint on the elbowgrease. Dig or fork as deeply as you can. Double dig if the subsoil is hard and intractable, but beware of bringing it to the surface. Your plants will appreciate the cool, deep root run, provided you let the soil settle for a few weeks before planting.

The Choice of Stone

I find geology almost as confusing as maths and tend to doze off when lectured to on the prehistoric origins of rocks. It's a shameful admission of a fault which I suspect is shared by many other gardeners. Rather like the amateur who wonders round an art gallery, I know what I like, even if I don't know why I like it. It is the appearance of a certain kind of rock and its ability to fit into my surroundings and among my plants that makes me choose it, rather than the fact that it is geologically correct for my geographical situation.

Having said that, I think it is certainly best to choose locally quarried stone for your garden, for not only will it fit naturally into the scheme of things, but it may also be much cheaper. The price you have to pay for cartage of stone may well equal the cost of the stone itself.

Of the many different types of stone, by far the commonest in rock-garden construction are limestone and sandstone. Limestone is likely to be of the Westmorland or Purbeck variety. The former is the more popular and the more expensive. It is usually sold as 'water-worn' and has been smoothly sculpted by flowing water into handsome shapes full of holes and runnels. Nowadays Westmorland stone is not easy to come by (which probably accounts for its high price), and much of that offered for sale has been taken from old rock gardens that have been dismantled by unappreciative owners. Thanks to this the stone is usually well furnished with moss, which adds to its lived-in look. Most plants grow well over and around limestone, though a few lime haters will not enjoy its proximity.

Sandstone is a friendly rock that fits easily into most gardens. Sussex sandstone is widely offered in and around that locality, and Kentish rag should be considered by gardeners further east.

I feel rather ashamed to confess that my own rock garden, built on a chalky slope in Hampshire, is constructed of a mixture of York and Forest of Dean sandstone. My excuse is that the house is surrounded by sandy-coloured shingle paths and a patio, and to have put limestone around this (for the rock-garden slope runs directly up from the house) would have seemed incongruous. Purists might have replaced all the gravel with limestone chippings – I chose the easy way out. Anyway, the plants seem to like their home, I like the effect, and the fact that the chalk is about 2ft (60cm) down is something I can turn a blind eye to so long as I avoid lime-hating plants.

As a Dalesman with a natural affinity to millstone grit, I have to admit that it can be unsympathetic in rock gardens, unless it is of a pleasing pastel shade. Granite is rather hard and unyielding too. Anyone who can get hold of that honey-coloured Cotswold stone should rejoice, for it looks good with or without plants clambering over it.

Tufa is a peculiar kind of rock that is formed by calcium carbonate deposits in certain streams and springs. In spite of its limy nature, many rock plants (including lime haters) seem to be able to grow happily over and in it (if holes are bored or chiselled out for pockets of soil and plant roots). I reckon it's best used in sink or trough gardens (see page 29) or a large lump drilled, planted and stood on its own as a feature. Used in the rock garden it is inclined to produce Mr Farrer's 'almond-pudding scheme' – a spotted mixture of soil and rock. But that's just my opinion. If you can get hold of decent-shaped pieces of tufa and fit them into your scheme, by all means use them. Moss will quickly colonise the nooks and crannies, making the rock look pleasantly antique, but it can also swamp the cushions of your plants. Weeds will easily get a foothold too.

When you've chosen the type of rock for your feature, shop around to get the best buy. Consult friends who have built rock gardens, and scour the Yellow Pages under 'Stone Merchants' to get the best buy. At the time of writing I have been quoted prices between £30 and £50 per ton of rock (including delivery). The price naturally varies from locality to locality, but almost everywhere Westmorland water-worn limestone is the most expensive, and local sandstone the cheapest.

The ton of sandstone used to build one of my rock banks consisted of around thirty pieces of rock. Three pieces were really hefty – about 2ft by 1½ft by 1½ft (60 by 45 by 45cm) – and when you are moving them on your own that's quite large enough! – and the rest weighed roughly half a hundredweight (25kg) apiece. I didn't weigh the larger pieces, but suspect they were a good two hundredweights (100kg) each, and levering them into a wheelbarrow was no easy task. If you ask for larger stones than these make sure you enlist the help of same slave labour when it comes to constructing the rock garden.

Too many small rocks will produce a spotty effect, and too many large ones an overpowering feature and the likelihood of a hernia. Try to ensure delivery of a mixed load comprising some large rocks which can be used as the heads of outcrops, some smaller ones which can back them up, and some large, flat rocks that can sit at the base or edge of the rock garden to give it a natural finish.

If money is in short supply I suggest you do without rock altogether. It's not so strange as it seems to grow alpines in a flat or raised bed topped with shingle or stone chippings and devoid of rock. Dwarf conifers can supply the missing height and, provided the soil is well drained, most plants will be idyllically happy in this rockless feature (see page 21 for details).

Building the Rock Garden

Advising gardeners how to construct a rock garden is comparable to entering an arena of lions. So many rules and regulations have been laid down in the past that one walks a tightrope, fearing that offence is always likely to be given to the purist unless the feature looks as if it has been lifted up and transported direct from the Pyrenees.

But rock outcrops occur naturally in very few British gardens, and however well you build them you will seldom fool any visitor into thinking that they were put there by a divine hand. Settle instead

Figure 1.1
An Artificial Rocky Outcrop

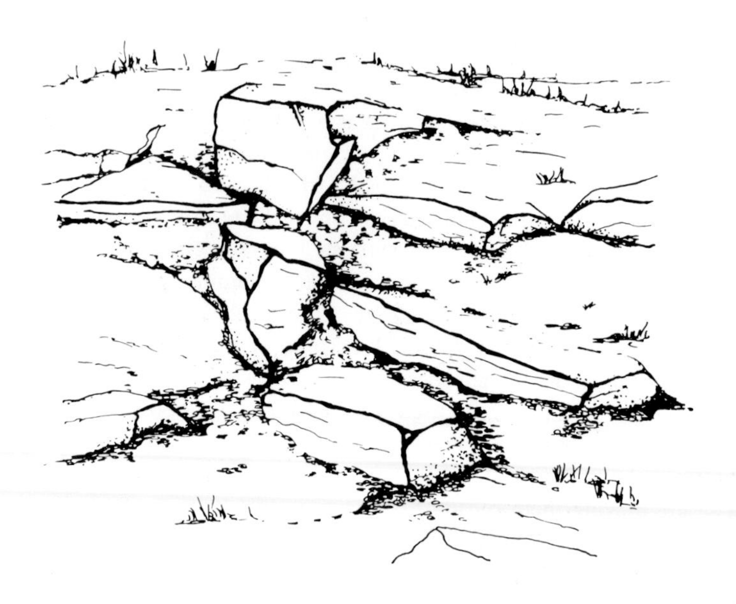

for an artificial 'natural' arrangement of rocks. Observe rocky outcrops in the countryside and on the coast. See how the layers within the rocks all run in the same direction – usually horizontally or sloping backwards slightly. Fissures tend to run vertically in a line and are seldom staggered as in brickwork. If you want to do your best with the rock, this is the effect you should emulate whether your garden is flat or sloping.

I seldom put pencil to graph paper when designing a garden, unless it is to show my wife the kind of thing I have in mind (this is not always a good idea, as changes frequently have to be made!). So it is with rock garden construction – you can plan as much as you like, but only when the rocks arrive and you set them in place will your inventive powers start to dictate the scheme.

A word or two on handling the rocks: try not to bash them more than you have to, for the fewer broken edges there are the more established the rocks will look. A sack barrow is the easiest means of transporting heavy stones, though I have used a wheelbarrow as a successful substitute by tipping it on its side, rolling a heavy rock into it and using the weight to right the barrow (fig. 1.2). Tip the rock into its prepared spot to avoid more manhandling than is absolutely necessary.

Ken Aslet, the late superintendent of the rock garden at Wisley, described what I find in practice is the most common-sense way of constructing a rock garden. He chose large pieces of rock to act as 'keystones' and having levered, pushed and pulled these into position

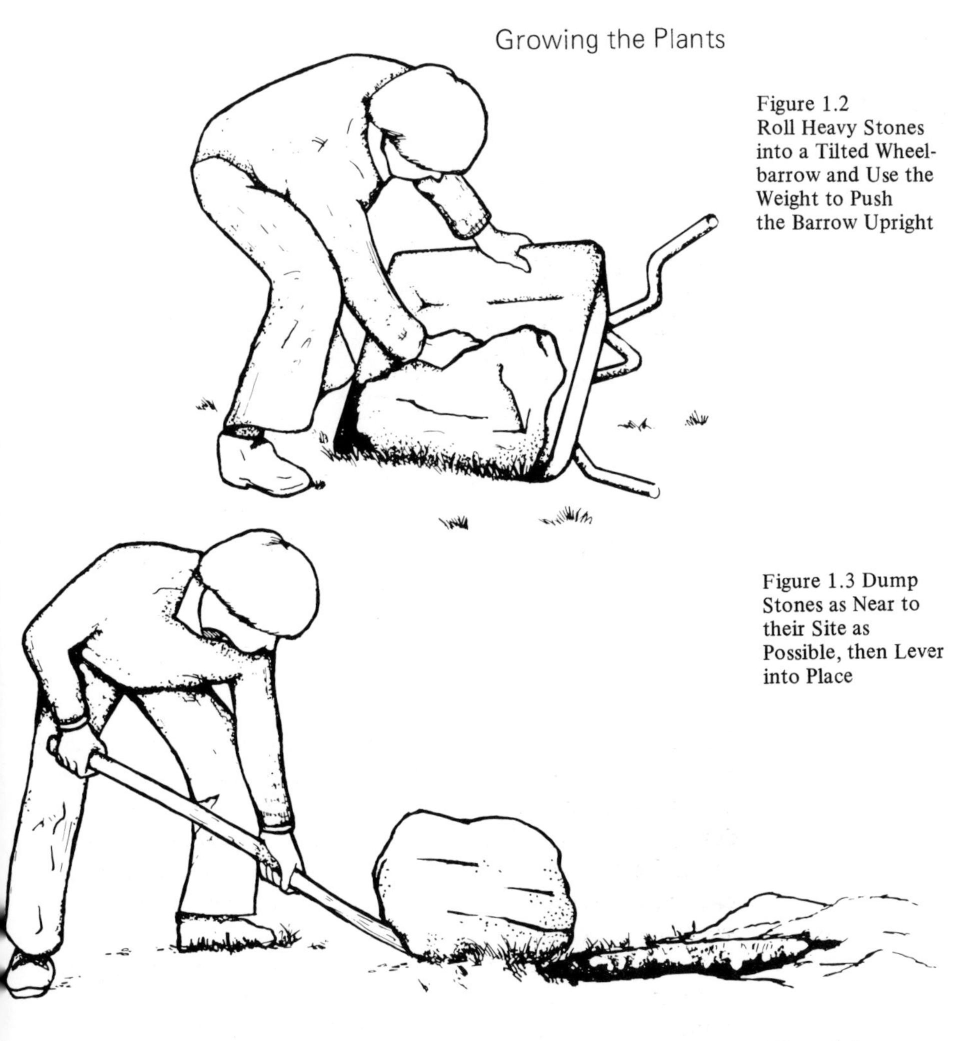

Figure 1.2
Roll Heavy Stones into a Tilted Wheelbarrow and Use the Weight to Push the Barrow Upright

Figure 1.3 Dump Stones as Near to their Site as Possible, then Lever into Place

Figure 1.4
Ram the Soil Firm with an Old Spade Handle or Large Dibber

he would work other stones around them. Each keystone should have a good face, and its slope will dictate the slope of all other stones. Don't place all your keystones at the same level. Have some at the top or back of the garden and some at the bottom or front.

Cutting out the soil with a spade, set each keystone to run slightly backwards into a slope, more steeply backwards into flat ground so that it seems to push out from the subsoil (fig. 1.3). Rain will be deflected from the upper surface to the plants' roots. Don't bury two-thirds of the stone as the old rock gardeners used to suggest (rock was cheaper and more readily available then), but do make sure that the rock is firmly bedded into the earth and that the soil is rammed under and around it (fig. 1.4). I use an old spade handle for this – the kind of tool that doubles as a dibber when brassicas are being planted on the vegetable plot.

With your keystone in place, select smaller rocks to arrange at the same tilt against its sides, above or below it, so that you create a natural-looking cluster. Flat stones can be part buried at the edge of the rock garden to act as kerbs. Always examine each piece of rock for its best face, and position this towards the front of the rock garden. Bury any newly broken edges so they are not seen, and where this is not possible, paint them with manure water (a spadeful of muck to a bucket of water).

Just how much soil you leave between each cluster of rocks is up to you, but I find that too thick a carpet of rocks is prohibitively expensive and cuts down the number of plants that can be fitted into the arrangement. The newly planted rock garden (see plate section) may look bare, but it will soon be colonised by carpeting plants that are prettier to look at than rock, and much cheaper too. Don't forget that the prime reason for building a rock garden is to grow plants that are shown off well among rocks. Too many gardeners seem to think that the plants should show off the rocks.

The contours of the rock garden are best made fairly slight where the feature is small. Only over larger areas will deep valleys and high ridges look at home. Avoid sheer faces of soil in any rock garden – they may crumble and will be washed away by rain. Leave sizeable sunny patches for carpeters, but remember to cater for shady-crevice lovers as well, creating niches between rocks where they will be happy to grow (fig. 1.5).

Plants like ramondas and haberleas and *Saxifraga longifolia* that are most easily established in vertical crevices, are best inserted as building progresses, for their root balls can be firmly wedged into the soil between neighbouring rocks – a difficult thing to do when two monstrous boulders are already in position.

If your soil is light and refuses to stay put in vertical crevices, wedge each aperture with turf. If the grass is turned inwards to the

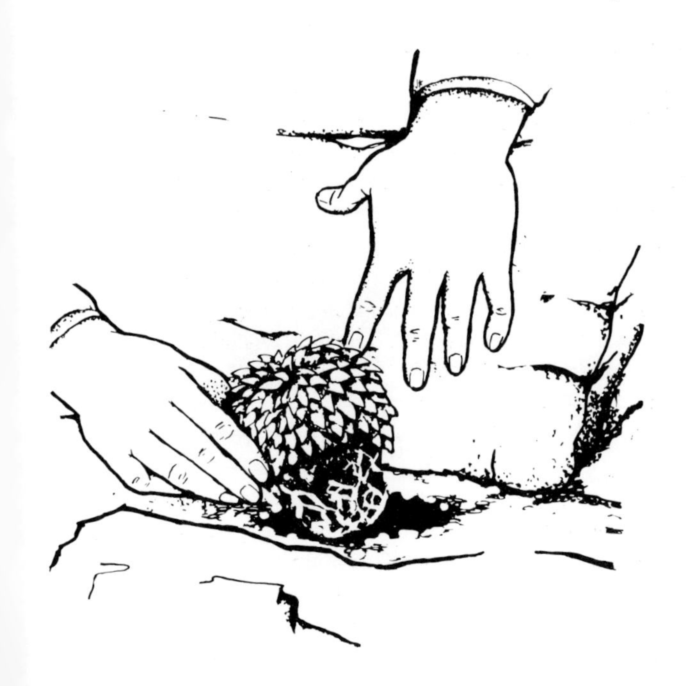

Figure 1.5 Crevice Plant Being Inserted During Rock Garden Construction

body of soil it will soon die off, leaving its fibrous roots to bind the soil together.

When your building work is completed, walk all over the rock garden from stone to stone to make sure that none of them wobble. Ram more soil into place around them if they are unstable. If your rock garden occupies a bank linking two parts of the garden, it will make life easier if you sneak some natural-looking stone steps into the scheme.

Trample all the soil with your feet once the job is finished, then rake the surface. Leave the earth to settle for a week or two before planting.

Rock Beds and Dry Walls

On flat sites and where space is limited a rock bed will enable you to grow a surprisingly large number of alpines (fig. 1.6). It will almost always be cheaper to construct than a rock garden for it does not demand much in the way of stonework. You can get away without using any if you wish, but I prefer to insert one or two shapely pieces of rock just to give the plants a taste of home and to remind the onlooker that he's observing alpines!

In our previous garden we made a 10ft by 3ft (3m by 1m) rock bed outside the back door against the east-facing wall of what we rather grandly called the conservatory. The soil was sharply drained Bagshot sand, and required no more than a good helping of peat

Figure 1.6
Rock Bed with
Raised Terraces

to put it in fine fettle for alpines. This sharp drainage meant that the bed did not need to be raised up. The edge (abutting a concrete path) was lined with reconstituted stone 'bricks' to make a neater finish.

On heavier soils it pays to raise the bed off the ground with two or three courses of bricks or broken flagstone pieces – not mortared together, just stacked in staggered courses with a slight backward tilt. Any gaps between the bricks or stones should be filled with soil. Unless you plan to include stepping-stones in your rock bed, make sure that you can reach all parts of it from the edges.

The enclosure is filled with a free-draining medium, say two parts garden soil to one part sphagnum peat and one part sharp sand or grit. The heavier the soil the more grit and chippings will be necessary. The importation of soil makes the feature more costly (unless you happen to be excavating a garden pond at the same time and can use the earth you remove in your rock bed!).

If a flat bed does not appeal to you, construct extra 'walls' across the top of the bed to make terraces of different depths. Trailing plants will look well cascading from one level to another.

Raised beds can even be built on concrete or flagstone patios, but they should be three to four bricks high and equipped with a drainage layer of broken bricks before the compost is put in place.

Plants such as sempervivums (houseleeks), encrusted saxifrages and other crevice dwellers can be inserted as the wall is built, and soil packed around their roots and in any other vacant fissures in the wall to prevent desiccation. When all the soil is in place insert your chosen rocks, taking as much care to position them with their strata running in the same direction as if you were making a fully fledged rock garden. Three or four decent-sized pieces will do for a bed 10ft (3m) long. When the job is finished, tread the bed thoroughly and rake it level. The soil should rest about 1in (25mm) below the rim of the bed to allow for a top-dressing of chippings (see page 34).

Scree Beds

It would seem a gross injustice to call screes 'nature's slag heaps', but that's what they resemble, even though they are cleaner and easier on the eye. At the foot of most mountain slopes you'll find gigantic piles of stone, usually broken into fairly fine pieces. Growing in this fiercely drained medium, where the barest quantities of soil and organic matter offer them sustenance, are some of the most beautiful and delicate alpines. Create a scree bed in a sunny part of your garden and you will find it easier to cultivate these plants that the alpine gardener describes as 'choice'.

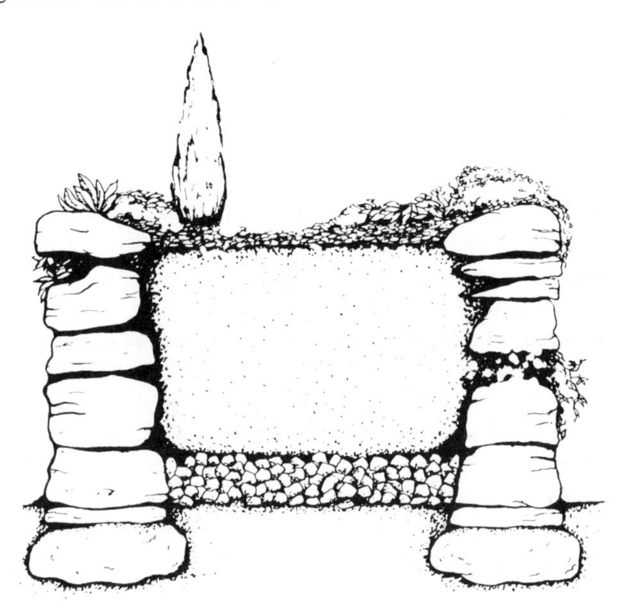

Figure 1.7
The Scree Bed Showing Foundations, Drainage Layer, Soil/Grit Mixture and Top-dressing of Chippings All Held in Place by Retaining Stone Walls

The scree bed is usually raised up from the ground (fig. 1.7). It can be excavated and made level with the soil, but on heavy ground there is a risk that it may act like a sump for drainage water unless it is situated at the top of a slope. On level ground over light, well-drained soil make the walls (as described under Rock Beds) 1ft (30cm) high; over heavier soils make them 1½ to 2ft (45 to 60cm) high. Walls this high are best stood on concrete foundations to improve their stability. Excavate a trench 9in (23cm) deep along the site of the wall and ram 3in (8cm) of brick rubble into the base. Top this with a 6-in (15-cm) layer of concrete tamped level with a wooden board. When this foundation is set (after two or three days) the first course of bricks or prepared stones should be mortared to the foundations, but drainage holes left at frequent intervals in the vertical crevices. If your stones vary in size, use the larger ones for the lower courses. As with rock beds, the walls should be given a slight inward tilt or 'batter', and all gappy joints should be filled with a 'greasy' or fibrous soil that will not fall out when it gets dry.

Railway sleepers held in place with stout posts can also be used to make scree beds (if you are not too particular about appearances!), but whatever you use keep the surface of your scree bed level for best results.

A 4 to 6-in (10 to 15-cm) layer of drainage material – broken bricks and rubble or flints – should be spread over the base of the bed, and the growing medium can then be put in place. Each alpinist has his own recipe for a scree medium, but the following is a good mixture which will provide your plants with sharp drainage and sufficient sustenance:

10 parts stone chippings (limestone, sandstone, flint or granite) or washed pea shingle
1 part good garden soil
1 part sphagnum peat or sifted leafmould
1 part coarse river sand

In areas of very low rainfall use five parts chippings instead of ten parts, so that moisture is retained longer. For lime-loving plants limestone chippings should be used, while plants preferring a more neutral soil should be given one of the alternatives. A 1-in (25-mm) top-dressing of the same chippings can be spread over the surface of the mixture to finish off the bed.

Several pieces of rock positioned in the bed will not only look good, but will also allow a cooler root run for the plants inserted against them.

Left to its own devices a scree bed will produce excellent results for several years, but it may then begin to look tired as the plants run short of nourishment. Counteract this by top-dressing the bed each autumn and spring with a light sifting of garden soil, sand and leafmould or peat mixed together in equal parts. The rain should take the mixture down through the stone chippings.

Watering, too, should not be neglected, for in this sharply drained medium plants can die in prolonged periods of drought during summer. In the Alps heavy rain showers are frequent, and there is no reason why they should not occur (via a hosepipe) on the scree garden at home.

A list of plants which are happiest when growing in scree beds is given on page 194.

Peat Beds

At the other end of the scale from the scree garden is the peat garden, where lovers of deep, acid, humus-rich earth and moist, shady conditions can be made to feel at home. The peat garden is undoubtedly a connoisseur's plot where rare petiolarid primulas, shortias, cassiopes, trilliums and the like can be cosseted and gloated over. It demands constant attention if it is not to become a home for vagrant weeds, and it should be constructed only by the patient plantsman who enjoys looking after a select band of plants. That said, a well-made and well-cared-for peat garden can offer unparalleled delights.

Only on neutral and acid soils are peat beds an unqualified success. They can be made over chalky soil if they are raised up a couple of feet (60 cm), if the bed is lined with polythene to prevent the invasion of chalky water, and if irrigation is always carried out with rainwater; but success cannot be guaranteed. If you garden on chalky soil, grow a few peat-lovers in large pots of a suitable ericaceous (lime-free) compost where they can more easily be protected from alkaline conditions. I don't like to be so discouraging, but it will be easier on you and your plants if you work *with* your conditions rather than *against* them.

Most peat-garden plants are shade-lovers, so choose either a bed against a north-facing wall or one shaded (but *not* overhung) by trees or shrubs. Situated directly beneath a tree canopy, your peat garden will suffer when the autumn harvest of leaves and drips begins to descend. Peat is the very devil to re-wet once it has dried out so, to save yourself labour and your plants the shock of drought, keep the bed out of full sun.

You're not trying to emulate a natural rocky outcrop when you build a peat bed, more a piece of heathland (uncolonised by grasses, gorse and heather!). A trip to the moors to observe their banks and terraces of peat will not come amiss, and you can then travel home as a qualified peat-garden architect.

Start off by cultivating the patch of earth that is to become the peat garden. Fork or dig it over to a depth of one spit (spade blade), and remove every last piece of weed root and stem. If you're in doubt as to the efficacy of your cultivations leave the bed alone for several weeks afterwards to see if more weeds emerge. Remove completely any that do. Perennial weeds with thick roots will wreak havoc in the peat bed, and are nigh on impossible to eradicate except by hand and with'touchweeders' such as Tumbleweed Gel (see page 15).

With weeds eradicated and the soil in good shape, the peat

enrichment can be added. You'll need to incorporate about as much peat into the top spit of soil as there is soil itself, so that you can create a 50/50 mixture of peat and earth. It's a lot, but it's what the plants like. Use *moist*, medium or coarse-grade sphagnum peat, and make sure that it is well mixed with the soil.

Figure 1.8
The Peat Bed with Terraces Held in Place by Peat Blocks and Logs

Once the soil has been enriched, the bed can be contoured, and the different levels or terraces supported by rows of peat blocks which are sloped slightly backwards and well anchored in the peaty earth (fig. 1.8). The blocks can be obtained from specialist nurseries or from firms specialising in peat cutting. If they are dry on delivery, soak them for several days in a large tank of water, weighing them down with bricks to prevent them from bobbing to the surface like corks. The peat blocks may be long and narrow or cube shaped (the larger they are, the less likely they are to dry out quickly), but whichever is the case they must be part buried in the earth to keep them moist. Peat-block walls whose stability is suspect can be skewered into place with bamboo canes. Never let the walls dry out – if they do they will shrink away from the earth and plant roots may be exposed to the air with fatal results.

If peat blocks are difficult to obtain in your area, use logs or even bricks to retain the terraces of peaty earth. The only disadvantage is that plant roots cannot penetrate them, as they can peat blocks, but otherwise they are useful substitutes. When construction is complete, lightly dust the surface of the soil with a lime-free general fertiliser such as blood, bone and fishmeal, and then mulch the area with a 2-in (5-cm) layer of moist, coarse peat. The plants need little in the way of additional food, but they will appreciate this snack when they are planted.

Avoid walking on the bed at all costs, except to firm it before planting. Lay stepping stones on the surface to make for easy access. Planting can be carried out in autumn or spring after firming by shuffling gently over the surface with your feet. Neaten the soil after planting by lightly forking it over, and from now on never let the peat blocks or the soil dry out. A sprinkler irrigation system will be a great labour-saver, but where it is out of the question a hosepipe fitted with a fine spray-head is essential.

Remove leaves, weeds, twigs and other detritus from the bed when it is noticed, and each spring add a further 2-in (5-cm) top-dressing of peat. Overcrowded plants can be lifted and divided in subsequent springs as necessary. A list of plants best grown in peat beds is given on page 195.

Space forces me to stop here, but if you are anxious to steep yourself in the ways of the peat garden you should seek out a copy of *The Peat Garden and its Plants* by Alfred Evans (Dent, 1974). It's a book that will make your mouth water!

Alpines in Patios

The patio crops up in every garden nowadays, even though the word was originally used in Spain to describe the courtyards adjacent to or enclosed by villas. Now it encompasses any patch of concrete or crazy paving that's used as a sitting-out area in bright weather. If all you possess is a patio, then you still have no excuse for neglecting alpines; choice species can be cultivated in sink gardens (see page 29), but many carpeting and cushion-forming plants will bask happily when planted in the crevices left between paving stones. Here they are assured of a cool, moist root run and a dry collar – the stone will prevent the sun from scorching the soil, and the plant's damp-detesting stems from rotting off. Provided the earth below the paving is well drained, it will keep the plants happy in winter too – plenty of sand or grit can be added to the soil before the patio is laid to speed up the removal of excess water.

Ideally you should plant up your patio during construction, for the plant roots can be set firmly in the earth before the flagstones are butted together. Cultivate, trample and rake level the soil before laying and treading on the stones to make sure they do not wobble, adding or subtracting soil until they are firm. Stagger the joints in the paving but leave gaps of 1½in (4cm) between them all to allow for plant growth and subsequent planting. Fill all the gaps with gritty soil to make an even, accident-free finish. Heavy soil should be lightened with plenty of sharp sand, grit and peat, and very sandy soil encouraged to hold more moisture by adding peat or well-rotted leafmould before the flags are laid.

Plant from pots between spring and autumn, and whatever the season water the plants well beforehand. Use a trowel and make sure that while the root ball is covered with earth which is firmed well around it, the stem is brought cleanly between the stones so that the stems rest flat on the upper surfaces. Water the plants in quite thoroughly after planting.

If your patio already exists then you'll have to adopt a different approach. If it's concrete, or the flagstones have been mortared into place and grouted, there's nothing for it but to carve out holes for your plants with a lump hammer and cold chisel. If you have to do this, buy yourself a cheap pair of plastic goggles to protect your eyes from flying chips of stone – I speak from bitter experience! Make the holes about 4in (10cm) across so that you can remove soil easily with a fern trowel (a long, narrow version of the ordinary trowel and a tool indispensable to the keen rock gardener.

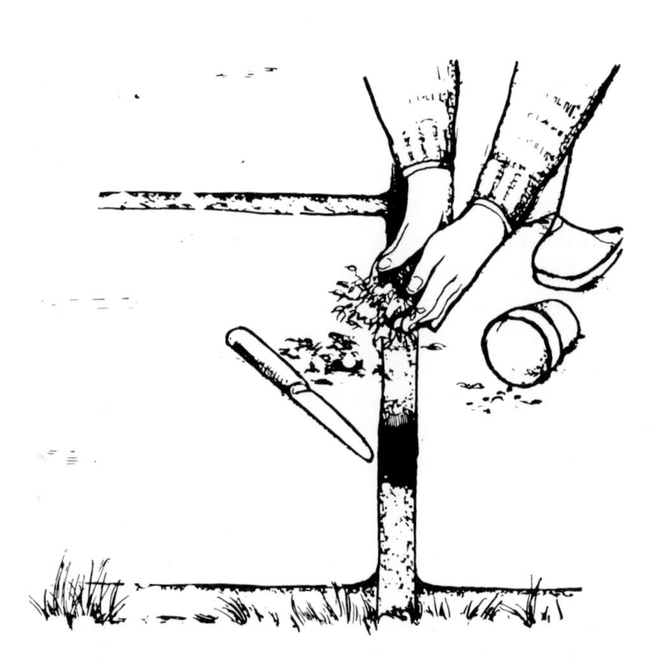

Figure 1.9 Planting Alpines in an Existing Patio

If there are soil-filled crevices between your paving stones then life is a little easier. Use an old kitchen knife to prise out the soil (Figure 1.9), then, after watering the plant and removing its pot, squash the root ball in your hand so that it is possible to squeeze it down through the fissure into contact with the soil (fig. 1.9). Replace the soil around the roots by trickling it in and ramming it into place with a ½-in (13-mm) piece of dowel or a fat pencil. Add sand or peat if the soil is poor. Again, a good watering will settle the roots into place.

Carpeting plants are best suited to this kind of culture, and some of them, such as chamomile and thyme, will even withstand being trampled on – the surest way of releasing their aroma. But do be gentle on all the plants in their first year of establishment; water them in dry spells and try not to crush them with a carelessly placed deck chair. A list of reliable paving plants can be found on page 195.

Trough and Sink Gardens

The interest shown in sink gardens today is tremendous – anyone with a little imagination is captivated by the miniature landscapes that can be created within the confines of a redundant piece of domestic equipment. Old stone sinks need no preparation other than a thorough scrub and the provision of drainage material, compost and plants; brown-glazed sinks don't look too bad untreated, but white porcelain sinks (not renowned for their intrinsic beauty) can easily be disguised to look as if they, too, are made of stone.

First, scrub the sink thoroughly to remove grease and dirt, dry it and throw away the plug, as good drainage is essential. At this stage it is a good idea to put the sink in its intended position – a sunny spot where it can be admired and appreciated without getting in the way. On patios and terraces sink gardens look a treat, provided they do not impede access. Moving the sink at this stage is not difficult; moving it later is likely to result in damage to the coating and to your internal organs. Stand the sink on four bricks to keep it clear of the ground, or build a brick plinth if you want to raise it up considerably – a must for wheelchair gardeners.

Paint the outer sides of the sink and 3in (8cm) over the inside rim with a bonding agent such as Unibond or Polybond – this will ensure that the 'stone' coating will stick to the porcelain (fig. 1.10).

The artificial stone coating I use consists of two parts sphagnum peat, one part sand and one part cement mixed together with water to form a firm mud that is patted on to the sticky surfaces of the sink. If the bonding agent dries before you pat on the ½-in (13-mm) coating of mixture (often called 'hypertufa') it won't stick very well, so make sure that it is tacky at the time of application (fig. 1.11).

Persevere if the mixture seems to fall off as soon as you slap it on; you'll soon get the knack of kneading it into place. When you've coated all the sink, leave it to dry for a couple of weeks before

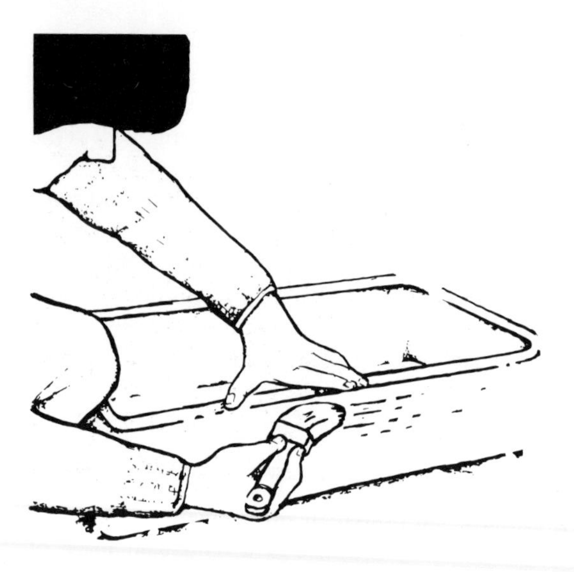

Figure 1.10
Sink Garden –
Painting with a
Bonding Agent

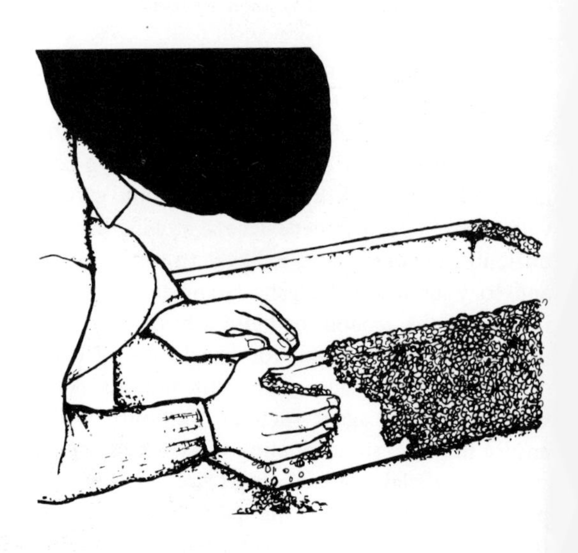

Figure 1.11
Sink Garden –
Patting on the
Hypertufa Coating

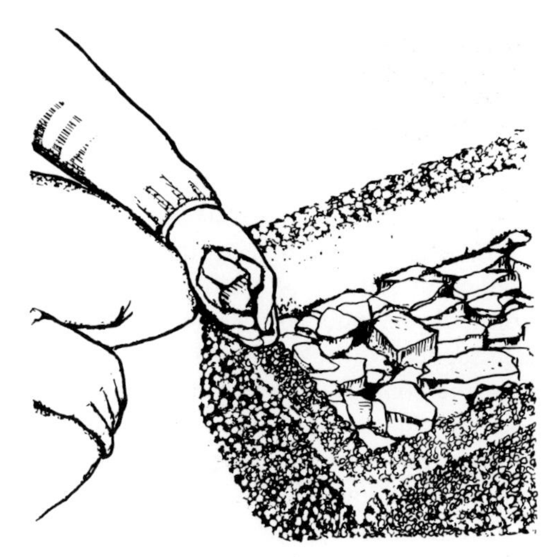

Figure 1.12 Sink Garden – Laying Broken Bricks or Crocks over the Base

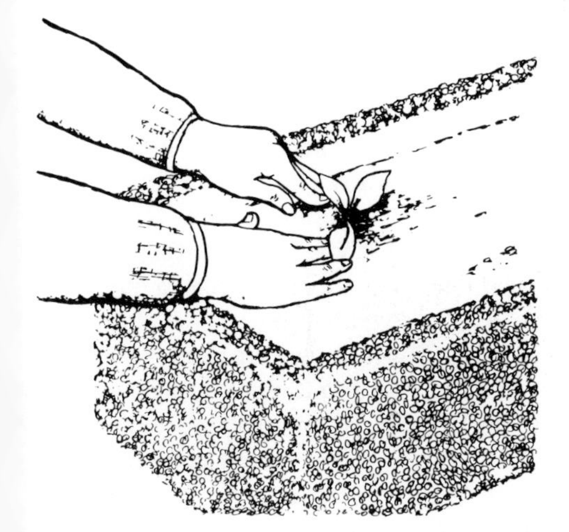

Figure 1.13 Planting the Alpines in Gritty Compost

Figure 1.14 The Completed Sink Garden

attempting to plant it up. After a day or two some folk scrub the half-dry mixture to produce a rougher surface; I've never bothered and the sink still looks reasonably stone-like, especially after a year or two when it's been colonised by mosses and lichens. Severe frosts may lift off patches of the stone coating in time, but it can easily be repaired by repeating the original process on naked areas.

The planting method is the same for all sinks and troughs, whether of natural or artificial stone. First cover the base with broken flower pots (if, like me, you are clumsy enough to have a large supply) or broken bricks and flints (fig. 1.12). Over this layer place the compost in which the plants will grow. I use John Innes No. 1 potting compost plus half its bulk of sharp sand for improved drainage. Fill to within 2in (5cm) of the rim with this mixture and lightly firm it down.

Sit one or two pieces of rock in position if you think they'll add to the picture (part bury them for stability and good looks) and then plant up with pot-grown alpines and one or two very dwarf conifers (fig. 1.12). Spring is the best time for this operation, and the surface of the compost should be coated with a ½ to 1-in (13 to 25-mm) layer of pea shingle or stone chippings when planting is completed (fig. 1.13). You'll be surprised how many plants you can get into a small sink – around 20 in one that measures 2ft by 1½ft (60cm by 45cm). The list on page 195 will give you some planting ideas, but don't be afraid to experiment a little. In a tiny garden a number of sinks (and a horse trough or two if you happen to have them lying around) can provide year-round delight if they are planted to flower in succession, or if each is given over to plants of a single genus – saxifrages, sempervivums, or androsaces. Silver-leaved alpines and those with white blooms will make a cool-coloured landscape, while fiery-flowered subjects with red-leaved sedums and sempervivums will make a richer scheme.

The possibilities are endless and always eyecatching, and only when a sink has really become overgrown and unsightly will it need to be emptied and replanted. Prolong its life by selecting at the outset plants which are not too rampant and which will be happy in each other's company. Confine trailers to the edges so that they can take up the vertical space outside the sink, and give plants that insist on a cool root run the spots next to the rocks. In winter the sink garden will need little care, other than the removal of fallen leaves, but in summer it will be dependent on you for water, so don't let it down. An occasional liquid feed will not come amiss – the plants' roots cannot search as far for nutrients as they can in open ground, so a little extra nourishment will help to keep them in good shape.

Alpine Lawns

The term 'alpine lawn' was originally coined by Clarence Elliott, one of the most informative and entertaining alpine plantsmen ever to put pen to paper. The fame of his Six Hills Nursery is perpetuated today by a number of fine plants that carry the 'Six Hills' epithet, and his son Joe Elliott continues the family tradition at his Broadwell Nursery in Gloucestershire. Clarence Elliott used the term 'alpine lawn' to describe a patch of well-drained ground that was planted up with dense, mat-forming alpines among which dwarf bulbs were planted to emerge and bloom in spring.

The feature caught on rapidly, but those who hastily gave over large areas to alpine lawns quickly realised that the technique demanded thorough ground preparation and careful plant selection if the lawn was not to turn into a rampant, weed-ridden carpet colonised by a few, all-pervading creepers, couch grass and ground elder.

To create an alpine lawn choose your site carefully. A patch of earth next to a patio or at the foot of the rock garden slope is suitable and, provided it is never soggy, it will quickly be colonised by a wide range of carpeters. But before you plant anything, do make sure that the ground is weed free, even if you have to wait a few months after cultivation to see if any aliens push through. Prepare your ground in spring or summer and plant it up in autumn when you know it's clean. For best results plant several plants of each species adjacent to one another. This will produce large drifts of leaf and flower colour, and will avoid the spotty look that results when single specimens of 50 different plants are mixed together. Planting distances may vary from species to species, depending on vigour, but on average 1ft (30cm) in either direction will allow for sideways growth while producing fairly rapid cover. A thin surface covering of sharp sand or fine grit will prevent mud splashing, while still allowing the plants to spread and root sideways with rapidity.

So that spring-flowering bulbs can be planted with ease in autumn, alpine lawns are best constructed in August or September. If planted in spring they may have romped away sufficiently by bulb-planting time to make it a tricky operation.

As with the plants, so with the bulbs; plant in sizeable drifts for best effect – a dozen of one variety here and two dozen of another variety there, rather than small numbers of everything scattered like hundreds and thousands on top of a cake.

In the early stages of growth it is important to give your alpine lawn every chance of succeeding. Water it thoroughly in dry weather with a lawn sprinkler, and remove any weeds as soon as they are seen. In spite of the fact that it is a lawn, walk over it with care – some of the carpeters may bruise easily, and if they're in flower you'll crush the life out of the blooms. For plants suitable for alpine lawns see page 195.

Planting

I've devoted a section to planting because it's an operation that is of prime importance, and because it's basically the same whether you are making a fully fledged rock garden, a sink garden or a peat bed. As to the best season for planting, spring and autumn are the traditional times, though summer planting from pots is possible provided you make sure the plants never dry out at the roots. Winter planting is inadvisable – the murky weather may well lead to rotting off.

Never plant anything unless you've got time to do it well. If you're in a hurry you'll just shoe-horn the plant into place, and it could be forgiven for dying rapidly in the belief that you didn't care about its welfare. Fortunately, all alpines are sold in pots, which means that planting gives them only a very small shock to their systems. Cut down on this by first of all watering them thoroughly an hour or two before you plant so that the root ball is moist. If it's dry at planting time it may well stay dry afterwards, however much water you pour on the surrounding soil. Remove any weeds from the pot.

Some gardeners (and I am one) like to dust the surface of the soil with bonemeal before planting. This is a fertiliser rich in phosphates, which promote root development without inducing vigorous growth. In spite of the fact that my old soil-science lecturer used to maintain that most phosphatic fertilisers were 'locked up' and made unavailable by the soil, I still like to add a sprinkling for alpines in the belief that if the plants have just a taste of it they'll feel better. My charity is probably totally misplaced, but I still dust a handful over each square yard (m^2) of ground.

When you're ready to plant, look at the size of the pot and then use a trowel to dig a hole that is 3in (8cm) larger all round. Slip the plant out of its pot (tear it off if it's paper or polythene; knock it off if it's plastic) and without disturbing the root ball sit it in the hole. Check that the surface of the root ball will rest level with the surface of the soil when planting is finished. Push back the soil, firming it well into the crevices around the root ball with your fingers or a piece of wooden dowel. Smooth off the surface of the soil when you've finished, and put the surface layer of gravel or peat in place. Now water the plant thoroughly.

A 1-in (25-mm) thick top-dressing of gravel or stone chippings on the rock garden (or peat on a peat bed) will not only improve the appearance of the feature but will also help the plants. It prevents mud-splashing, keeps down weeds and keeps in moisture during summer. It also protects the 'necks' of alpines from excessive moisture.

Watering during the first spring and summer after planting is especially important; neglect it and your treasures will turn crisp and brown. When you think that water is necessary (scrape away some of the gravel or chippings to observe the state of the soil) apply it generously. Leave a sprinkler running over the area for at least an hour so that the earth is well drenched. A light spray over will do no good at all.

Alpines under Glass

Sooner or later any gardener deeply bitten by the alpine bug will want to grow those mountain plants that cannot survive outdoors during soggy British winters. This is not due to any cussedness on the part of the enthusiast, it's simply that the damp-hating, high-altitude alpines are so captivating that it seems worth going to any lengths to cultivate them successfully. Among this vast collection of choice plants are the cushion-forming androsaces and dionysias, the feathery paraquilegias and species of corydalis, douglasia, phyteuma and tricky campanulas. The list is endless and so is the enjoyment.

Two sound little books, sadly now out of print and in part out of date, have been written on growing alpines in special greenhouses devoted to their well-being. *The Alpine House And Its Plants*, by Stuart Boothman (Rush & Warwick, 1938) and *Alpine House Culture for Amateurs*, by Gwendolyn Anley (Country Life, 1938) are both worth reading if you can pick them up in a secondhand bookshop or local library. The newly reprinted *Collector's Alpines* by Royton E. Heath (Collingridge, 1964) is an invaluable bible for the devoted enthusiast.

I have little room here to delve deeply into the construction and setting up of a fully fledged alpine house, but I will say that these choice plants can be grown well much more easily and in many more ways than some folk would have you believe. True enough, there are a few species which defy all attempts at cultivation, but there are many more that are willing to oblige you – if you oblige them. Most high-alpine plants will thrive if you give them what they need in the way of compost, water and climate. This is easily said, but not always easily done. Find out where they grow, what they grow in, and what the weather is like, and do your best to imitate these conditions. If you succeed to even a modest degree, the chances are that you will grow your alpines well.

Choice alpines are grown in greenhouses and frames to give them protection from icy winds, damaging winter rains and fog. They do not need heat (except to reduce the severity of hard frosts), but they do need excellent light and excellent ventilation. The traditional alpine house is a specially manufactured greenhouse with continuous vents on both sides of the ridge and on both eaves, plus a few vents below the benches as well. The experts say that the house should run north/south, but in my experience an east/west

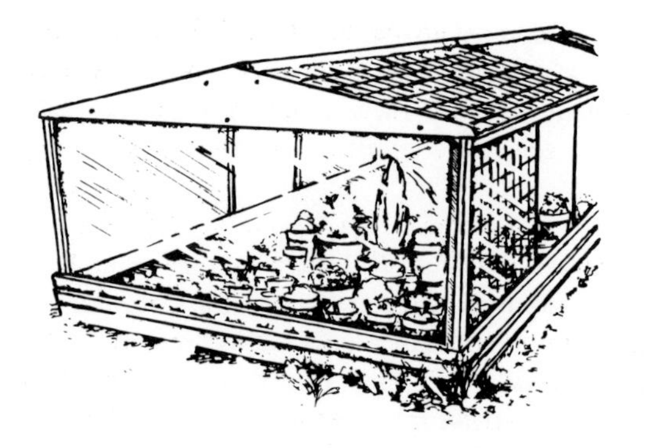

Figure 1.15 A Double-span Frame for Alpines – Shading Material in Place

house has never prevented a good gardener from succeeding with alpines. Wooden houses are better than aluminium ones as far as growing alpines is concerned, for there is less of a problem with condensation drip, which can easily spoil a good cushion plant. If you can afford a purpose-built alpine house, and have the energy and enthusiasm to look after it, then invest in one, for it will show off your plants to perfection. Alternatively, grow your plants in a well-lit garden frame (fig. 1.15).

In my own garden I compromise. A south-west-facing lean-to conservatory with good, but by no means continuous, ventilation acts as my alpine house from January to April. It is ventilated well at all times during the day (except during foggy or bitterly cold weather), and the door plays a vital part in ensuring a through-flow of air! If a breeze is blowing, always ventilate on the leeward side, if you have the option, to avoid draughts. In winter the house is shut down at night, but in spring a 'crack of air' is left on to avoid a stagnant atmosphere. For the rest of the year the plants are transferred to a double-span, shingle-filled frame on the east side of the house. There's no reason for this aspect other than convenience. Here they are part plunged in the shingle, checked daily for water and health, and shaded from bright summer sun with green, close-weave, plastic netting. The sides of the frame remain open to allow free air circulation whenever the weather is remotely clement, and in summer only the shading stays in place – the glass is stacked in an out-of-the-way corner. Rain in summer perks the plants up rather than doing them any harm.

By part plunging the pots the rate at which they dry out is slowed down. Plunge them totally if you like, but you'll have to add peat to the shingle so that it stays in place when you lift the pot out to check it for water. The standard recommendation for growing the plants in an alpine house is that they should be plunged on shingle-filled benches. This is desirable but by no means essential, unless the plants are to stay in the house through the summer, in which case they will enjoy its cooling influence. Shingle-filled benches need to be very sturdy and are expensive to construct. My plants stand on slatted wooden benches and the floor of the house is covered with shingle which can be watered to moisten the atmosphere on spring days.

There is much argument about the desirability of heating in an alpine house or frame. I find it useful. In severe winters such as that endured in 1981/82, electric tubular heaters set to switch on at around freezing point would have saved several plants that perished – thanks to the fact that I'd just moved house and had not equipped my temporary alpine frame with heat! As well as keeping the plants dry, that natural blanket of snow also prevents temperatures from falling to extremely low levels – something that a little electric heating can simulate.

For those gardeners who gasp at the very thought of electric heating, let me offer a word or two of reassurance. At the time of writing the relative costs of electricity and paraffin mean that a thermostatically controlled electric heating system (set to switch on only when frost threatens) is far cheaper than a paraffin heater which, apart from running constantly whether it is needed or not, gives out vast quantities of plant-poisoning fumes and water vapour. Don't use a paraffin heater to keep the frost off your alpines – it will do them more harm than good – but think also about replacing it with an electric heater in any other greenhouse that is heated to around 7°C (45°F). The cleanliness, economy and trouble-free operation of electric heaters makes them by far the best bet.

A little heat might be important in winter, but a little shade is equally important in summer, simply to keep temperatures down and prevent the plants from scorching. Traditional shading materials are often too thick for alpines, causing them to become drawn and spindly. If you have the wherewithal you can buy external blinds for your greenhouse or frame (always better than the internal type, for they intercept the sun before it hits the glass). The type made of plastic 'straws' or laths will usually admit a dappled light, and if the shade cast is too dense you may be able to remove every other lath – if you have the patience.

Plastic, close-weave netting is the material I use. It's green, not too unsightly and not dense enough to cut down too much light. Rig it up so that it can quickly be taken down on dull days when every scrap of light is needed by the plants.

Pots, Potting and Composts

I've indicated in the A-Z section which plants can be grown successfully in pots in an alpine house or frame. Some gardeners will bung absolutely everything in a pot, but I find this time consuming, space consuming and often unsatisfactory from the plant's point of view. It's far better, surely, to grow the plant where it flourishes best and needs least fussing over, so the only plants I recommend for pot or pan culture are those which either cannot survive outdoors, or which are better seen and admired under glass, where they can be protected from the bad weather which might spoil their early blooms. Only occasionally will I make an exception to this general rule, and that is when a plant is so spectacular that it adds tremendously to the alpine house display; and even then I'll grow it in a pot only if it is happy when so confined.

There are always arguments about clay and plastic pots and I suspect there always will be. Master your techniques of cultivation and you can grow most alpines in either, but I do personally favour clay pots for practical as well as aesthetic reasons. Their big advantage is their porosity – they can be plunged in sand or a mixture of gravel and peat in a garden frame and will dry out only very slowly, thanks to the fact that they can absorb moisture from the surrounding medium. With those plants that resent a really wet medium this is a great boon, for only the plunge material need be watered. Pots are fine for deep-rooted subjects, but pans (shallower containers) are suitable for most ordinary alpines, which spread sideways more eagerly than they root downwards. Whichever kind of pot or pan you use, do make sure that it's washed thoroughly before use. New clay pots and pans should be soaked for a day before being used so that any impurities are dissolved and the terracotta does not absorb moisture from the compost.

If there's a greater controversy than that which rages over pots, it's the one over composts. Each grower has his or her own favourite recipes, which rank with the bottles in a chemist's shop for complicatedness of composition. Again, perhaps due to my inherent lazinesss, I always look for an easy way out, but a successful one. Some growers keep bins of loam, peat, sand, leafmould, crushed tufa, brick dust and the like under the benches in their potting sheds. I've always wished I could be as organised, but so far I've never managed it and have to rely on easier methods. All the composts I use are based on John Innes potting compost – usually No. 1 but sometimes Nos 2 and 3. For lime haters I use John Innes ericaceous compost as a basis because it's lime free. In both types of compost there is a mixture of loam, peat and sand which more fastidious gardeners might go to the bother of mixing themselves. The compost is also sterilised and so free of harmful pests and diseases. I'd rather locate a good brand of John Innes (and be warned, there are some duff ones around) and add what I wanted to it. This varies from really sharp grit or coarse sand (either is suitable) to limestone chippings, peat or leafmould.

The sand or grit should be really sharp with particles ranging in size from about $\frac{1}{16}$ to $\frac{3}{16}$ in (2 to 6mm) – never use that bright-yellow builders' sand which is too mud-making and full of impurities. Leafmould (preferably oak or beech) is shunned by some growers for good reason: either they have difficulty in getting hold of it, or the stuff they have used encouraged root aphids in their plants. I recommend its use occasionally because I've found that its texture is just what some composts need, but as it is unsterilised and may occasionally give rise to problems, you must decide whether the risk is worthwhile. Peat – of a reasonably coarse sphagnum-moss grade – is a good, sterile substitute and a fine additive on its own. It aerates compost at the same time as helping it to hold on to moisture. Limestone chippings really cheer up lime-lovers, but if you can't get hold of them a sprinkling of ground limestone in the compost will do almost as well.

Throughout the A-Z section I've described the compost preferred by each pot-grown alpine. I apologise if it seems absurdly repetitious, but it seemed better to me that the information should be right where you need it rather than on another page which is cross-referenced. The recipes are not too fussy in most instances, and are based on what you might call a 'feel for compost'. Commonsense plays a great part here. When the compost has been mixed, feel it in your hands – rub it and squeeze it. Does it feel as though the plant will enjoy it? Will it be sufficiently sharply drained for those plants that need brisk drainage? Will it hold on to moisture well without becoming compacted for those plants that like a humus-rich medium? Fill a pot up with the stuff, firm it as if you were potting a plant and then water it. Notice whether or not the water runs away quickly – it always should. Watch it over a few days to see if it retains its moisture (if you have the time and the patience). Eventually you'll get the feel of composts and be able to tell if a particular mixture is right for a particular plant. This may sound like a vague and magical way of carrying on, but it's a better system than one based dogmatically on a teaspoon of this and a potful of that.

Alpines are almost always bought in small pots nowadays, which makes transferring them to the garden or into permanent pots a much easier and less risky job than when they have been lifted from open ground. You can pot them up at any time, except in the middle of winter when most of them would rather rest. Select your container (which should allow enough room for growth without being too large) and sit a layer of drainage material in the base. Broken flowerpots yield 'crocks' which are especially good for this purpose, but if you are never clumsy enough to have a good stock you'll have to rely on coarse gravel. Spread a layer of the gravel, or several pieces of broken flowerpot (concave side downwards) over the drainage hole or holes. A ½-in (13-mm) layer is quite sufficient, no matter what you might read to the contrary. Some old gardeners used to half-fill their flowerpots with drainage material. This is quite unnecessary – once the water has left the compost it has gone, whether it drains through ½in (13mm) or 3in (8cm) of crocks.

Water any plant thoroughly an hour before it is due to be potted. Most alpines – especially the cushion formers – are potted singly, but there is no reason why several plants should not be potted into one large container to give a more impressive display more quickly. Proceed as for potting a single specimen, but space the plants apart equally in the pan. To pot a single plant, put a little of the new compost into the base of the new pot and then tap the plant from its existing container, keeping as much compost around the roots as possible and only scraping away a little around the 'collar' of the plant to remove mosses and liverworts that might be present. Sit the plant in place in its new pot and check that the surface of the root ball rests about 1in (25mm) below the rim. Fill around it with compost, gently pushing this into place with your fingers. Continue until the compost is level with the top of the root ball. Now spread a layer of ¼-in (6-mm) grit over the surface of the soil to a depth of ½ to ¾in (13 to 20mm). Not only does this make the plant look smarter, but it has the vital function of keeping its collar dry and stopping any rotting. It also prevents mud splashing. Some plants like flat pieces of rock between their cushions and the compost; these are indicated in the text. Peat lovers should be top-dressed not with chippings but with flaked bark or sieved leafmould. Water the plant in immediately, and plunge it in a frame. Avoid exposing it to bright sunshine for a few days, even if it is a sun-lover. Until it recovers from the shock of transplanting it may wilt and scorch if placed in the sun.

'Double potting' can be practised on some of the more difficult plants such as dionysias. Here the plants are potted as described in small clay containers which are then sunk to their rims in larger pots of sharp sand. Any water is applied to the sand in the larger pot and makes its way through the smaller porous pot, keeping the compost around the plant's roots evenly moist but never soggy.

Repotting will be necessary at some time – either annually with fast growers, or every two or three years with those that grow more slowly. Recommendations are given in the A-Z section. Don't be mean to your plants. Certainly avoid disturbing them if they resent it, but top-dress any slow-growing plants in spring to give them a freshen up. Remove the gravel, scrape away ½in (13mm) or so of the compost and replace it with fresh before you put back the gravel top-dressing. There are one or two plants which will resent even this, but most will be grateful.

Watering

I'm sure you've heard that more pot plants die as a result of over-watering than anything else. Some alpines can die from under-watering too. It's a fact that the mastering of watering is the most important factor in growing alpines well, and yet it's another skill that relies a lot on instinct. During their growing season (usually from spring to early autumn) most alpines need a supply of water at all times so that they do not wilt. This does not mean that they should be kept soggy; nor that they should be allowed to dry out completely between waterings. It means rather that they should be watered well, and then watered again just before the compost dries so that growth is not checked.

I always water from above, and have not yet killed anything except one or two dionysias by doing so. The pot is filled to its rim and allowed to drain, and then watered again if I think it has not had enough. You can't do the trick of feeling the compost with alpines because they are covered with a top-dressing of grit, so you'll have to get used to weighing the pots in your hand individually and deciding by the weight if the compost is wet or dry. You'll soon get used to the idea and will be able to prevent any plant from wilting due to over or underwatering. Check your plants once a day, winter and summer.

Some growers insist on watering all their alpines from below – standing them in deep trays of water until they have taken up what they want. I admire their patience, but, except with a few tricky plants, have never found the technique to be essential.

Use rainwater if you have some means of collecting it; use tap-water if you haven't. If your water is lime and chlorine laden and you want to grow lime-haters and really choice alpines, then you'll have to rig up some system of rainwater collection. Use a water butt kept inside the alpine house, or at least covered with a lid to keep out leaves and muck. I water my alpines with rainwater while the supply is plentiful, but in summer they have to get used to tapwater.

Not many alpines need feeding. Those that do are indicated as such in the text, and can be given a good watering with a proprietary liquid fertiliser diluted according to the manufacturer's instructions. Liquid feeds based on seaweed are particularly good.

In winter most pot-grown alpines need to be kept *almost* dry. This is a state of affairs that's difficult to define. At no time should the foliage be allowed to wilt or become flaccid, but similarly the compost should never be wetter than a tightly wrung-out flannel. Your plants will recover from wilting due to dryness at the roots, but one mistimed watering may kill them, so err on the side of dry-ness. Keep all water off the foliage at this time of year or rot may quickly set in and kill the plant.

Pruning and Tidying

Detailed pruning instructions are given where necessary in the A-Z section, but several general rules apply: Always prune back to just above a leaf or visible bud; if you don't the stem will die back to such a point and probably beyond, allowing in fungal infection. Cut out any dead wood as soon as it is seen, again making your cut just above a bud or leaf joint. Remove dead and dying foliage immediately, before any infections have time to take hold. In winter, especially, such hygiene is vitally important.

Unless plants form decorative seedheads (or unless you are hoping to collect the seeds for propagation) snip off all conspicuous flower-heads as soon as they have faded. Not only is this in the best interests of the plant's health, but it might also induce a second flush of flowers later in the season.

Labelling

Always label your plants to show the name of the plant, the date of its acquisition and the source of supply. That way you'll know how old it is and whether or not the nurseryman supplied you with a good plant when you see how it performs in the future (provided you look after it properly). I find that metal Hartley labels are best – they can be written on in pencil or indelible ink, either of which lasts for ages. What's more, the labels never become brittle like plastic ones, and they can be scrubbed clean and used again should the plant meet its demise.

This type of label is also useful on the rock garden. I hate seeing rows of white labels sticking up like gravestones behind plants; they distract the eye from the alpine scene that has been created, and they are usually flung around with gay abandon by birds so that nothing is labelled correctly. These metal types can be pushed right down into the soil so that just ¼ or ½in (6 or 13mm) is left showing. When you forget what the plant is, the label can be pulled out to check and then pushed back in again.

A Rock Gardener's Calendar

Spring

Plant new alpines in the garden and in containers.

Treat deep-rooted weeds individually with a spot weedkiller such as Tumbleweed Gel (glyphosate). Pull out annual weeds.

Apply a light dusting of blood, bone and fishmeal to the rock garden or bed so that it can be washed in by rain.

Re-plant (and reconstruct if necessary) any patches of the rock garden that are overgrown or unsightly.

Start to water more regularly alpines growing in pots.

Repot alpines that have outgrown their containers (unless this is

better done after flowering – see A-Z section for specific details).
Ventilate thoroughly all frames and greenhouses that are sheltering alpine plants.
Construct sink gardens and new rock features on well-prepared soil.
Sow seeds of alpines in pots placed in a frame or propagator.
Protect plants from slug attack.
Protect tender plants from damaging frosts.
Divide clump-forming alpines.
Rig up shading material on very bright spring days when the temperature looks like rising dramatically.
Clear plants of any dead leaves and stems.

Summer

Apply water to parts of the rock garden or scree beds that are dry.
Water sink gardens and pot-grown alpines as frequently as necessary.
Continue to weed as necessary.
Take cuttings of alpines and root them in a frame or propagator.
Remove faded flowers unless they are decorative or seed producing.
Pot up alpines bought in from a nursery.
Repot those alpines best moved after flowering.
Continue to plant alpines outdoors except in baking weather.
Plant or pot up autumn-flowering bulbs.
Shade all plants under glass and ventilate well night and day.
Pot-grown alpines that need to be dried off after flowering should be placed in a sunny frame. Shade-loving alpines that have finished flowering should be plunged in a shady frame and sprayed daily with rainwater.

Autumn

Start to water more carefully any plants grown in pots.
Remove faded leaves and flowers from plants as soon as they are seen.
Divide clump-forming alpines.
Prepare the site for new rock gardens to be planted up in spring.
Ventilate freely any greenhouses and frames containing alpines, except on very windy or foggy days.
Plant or pot up spring-flowering bulbs.
Protect plants from slug attack.
Cover damp-detesting alpines with sheets of glass or cloches if they are being grown in the open.

Winter

Take great care not to overwater pot-grown alpines. Keep water away from the foliage.
Ventilate cautiously at all times.

Arrange for heaters to be switched on in the alpine house if the temperature falls below freezing point.
Remove all faded foliage from pot-grown plants.
Clean out the alpine house or frame.
Wash pots and pans ready for spring planting.
Prepare composts for spring planting.

Pests and Diseases

Grow your alpines well and you'll have few serious pest or disease problems, other than those which occur regardless of the health of your plants. The following are the most common troubles, all of which are best dealt with as soon as they are seen, to minimise damage and the likelihood of further outbreaks.

Always treat pesticides with respect. Use them strictly in accordance with the manufacturer's instructions, and store them out of the reach of children and pets. Surplus mixture can be disposed of down the lavatory, but try to avoid mixing more of the solution than you can use at one go. Don't store ready-mixed chemicals for more than a few days. Apply sprays on still days when there is little likelihood of rain.

Ants

These insects farm greenfly and can disturb plant roots on the rock garden when they make nests. Control them either by laying poisoned ant baits (in gel form), by dusting with an antkiller powder, by pouring on boiling water (if you can do this without harming plants) or by spraying the area with an aerosol insecticide such as Dethlac, which kills all insects that cross the sprayed patch.

Aphids

Greenfly and blackfly are fond of most plants. Spray at the first sign of attack with an insecticide based on pirimicarb (such as ICI Rapid Greenfly Killer). This is harmless to beneficial insects such as ladybirds, lacewings and bees.

Carnation fly

The maggots of this fly tunnel into stems, leaves and roots. Prevent attack by dusting the soil with insecticidal granules based on diazinon (e.g. May & Baker's Soil Insecticide Granules or Fison's Combat Soil Insecticide). Alternatively, spray attacked plants with an insecticide containing gamma-HCH (e.g. Murphy Lindex Garden Spray).

Caterpillars

Small infestations can be picked off by hand; larger outbreaks can be sprayed with ICI Sybol 2 or an insecticide based on fenitrothion.

Grey mould

This is a fungus disease also known as botrytis. It causes plant tissue to rot, and it covers infected areas with a greyish fur. Control by being hygienic and removing any faded foliage and flowers before rot sets in. Ventilate well whenever possible to allow good air circulation, which will lessen the likelihood of attack. Spray with a fungicide based on benomyl (e.g. Benlate).

Mildew

This is another fungus disease that thrives in similar conditions to grey mould. This time a white, powdery film is formed. Preventive measures are the same as for grey mould, and benomyl is again useful.

Red spider mites

These pin-prick-sized mites attack plants, sucking sap and spinning fine webs, so reducing the strength of the plant, as well as the strength of the light that reaches it. Control them by spraying with an insecticide based on malathion or fenitrothion. If repeated outbreaks occur, vary the spray used, or the pest may build up a resistance to one particular chemical.

Root aphids

Plants lacking in vigour or wilting may be suffering from an attack of root aphid – a greyish aphid that can be seen when the plant is tapped from its pot. Protective white wool may also be visible. Water infested plants with a diluted solution of malathion.

Scale insects

Like tiny barnacles in appearance and dark brown in colour, scale insects fasten themselves to stems and leaves where they suck sap and secrete honeydew. They can be killed by spraying with an insecticide based on permethrin (e.g. Bio Flydown).

Slugs and snails

Sprinkle mini-pellets among susceptible rock plants and renew these every few weeks in showery weather. Yoghurt pots sunk into the ground and filled with beer make excellent traps.

Vine weevil

White grubs eat the roots of plants, causing wilting and eventual death. Prevent attacks by using only sterilised compost. Water with an insecticide based on gamma-HCH (e.g. Murphy Lindex Garden Spray).

Whitefly

The tiny, white, v-shaped flies sit on the undersides of leaves and fly around in circles when disturbed. Fumigate with a smoke especially formulated to control whitefly (the pest usually occurs under glass). Repeat at recommended intervals to control emerging nymphs.

Woodlice

These eat the roots of alpines and cause death. Prevent attacks by using sterilised composts. Use potato halves as traps among infested rock plants. Pot-grown plants can be watered with a diluted solution of gamma-HCH (e.g. Murphy Lindex Garden Spray).

All pests and diseases that occur under glass can usually be more safely controlled by fumigating with smoke canisters than they can by spraying (which may produce unwanted humidity in autumn and winter). Check all chemicals and fumigants for their suitability to the particular plant or plants to be treated. If you are in doubt as to the safety of a particular product, consult the manufacturer, or test the substance on one plant before you use it on the rest.

Propagation

Plant propagation is a fascinating and rewarding craft, and one that every true gardener enjoys. Not only does it give tremendous satisfaction to produce a new group of plants, but it also saves money. All the plants described in this book can be propagated in one of the following ways: seeds, cuttings, layers or divisions.

Seeds

Seed sowing is certainly the cheapest and often the easiest way of producing a large number of plants, and it may be the only way of increasing those alpines which die once they have flowered – plants which are described as being 'monocarpic'. Plants which are difficult to obtain through the nursery trade are often offered in seed form by specialist societies, so a knowledge of their germination requirements is essential.

Most seeds are sown in spring so that they can grow uninterrupted through the summer, making decent-sized plants that are capable of surviving their first winter. Some seeds, though, lost their viability very quickly, and must be sown as soon as ripe. If you are collecting seeds from your own plants, keep a close watch on the seedpods and remove them just before they shed their seeds. All seeds can be stored for long or short periods in airtight containers placed in a refrigerator.

Seeds of alpines are almost always sown in pots, unless they are so easy that scattering them on the part of the rock garden they are to occupy is an effective way of ensuring establishment. Never use garden soil as a seed compost in pots; use one of the proprietary brands which will have been sterilised. Throughout the text I have recommended peat-based seed composts where necessary, and said simply 'seed compost' where brands such as John Innes are equally good. This is a personal preference based on successes and failures, and it cannot be denied that there are advantages and disadvantages with both types of compost. Peat composts in particular must never be allowed to dry out. Not only will dryness interrupt the development of the seedlings, but it will also cause the compost to shrink from the sides of the pot, making future watering difficult. John Innes compost can turn sour if it has to sit in a pot for two years or so while slow seeds germinate.

Choose pots at least 4in (10cm) in diameter for seed sowing as smaller containers tend to dry out too quickly. Where plastic pots are used (and I always use them for seed sowing) a layer of drainage material should not be necessary for the pot has a good supply of holes in the base. A layer of gravel or broken flowerpots should be placed in the bottom of single-holed clay pots.

Figure 1.16 Sowing Alpine Seeds in a Pot

Fill the container to the rim with compost, then tamp it down with the base of another flowerpot so that it rests ½in (13mm) below the rim. Sow the seeds thinly on the surface by tipping them into your left hand which is held in a lightly cupped shape, then tapping the side of your left hand with the fingers of your right so that the seeds fall on to the compost. Hold your hand about 6in (15cm) above the compost and the seeds will bounce a little and be distributed evenly (fig. 1.16). Very fine seeds need no further covering, but most seeds should be covered with a little sieved compost – stop sieving as soon as they disappear from view. Seeds which are slow to germinate can be covered with a light sifting of coarse sand to prevent the growth of mosses and liverworts. Label each pot as you sow, and then water it by immersing it in a tray of water. Remove the pot as soon as the surface of the compost has darkened with moisture. All future waterings are best given by this method to prevent the disturbance of the seeds which would result if a watering can were used.

Most alpine seeds need no artificial heat to persuade them to germinate, especially if sown in spring when the natural rise of outdoor temperatures will be enough to speed them into action. The pots can be stood in a garden frame which is shaded from bright sunshine and where they can be closely watched to make sure they do not dry out. If the pots can be plunged in gravel or sand within the frame so much the better, for this will allow the compost to dry out much more slowly.

Figure 1.17 Small Frame for Germinating Alpine Seeds

Those seeds which need a little heat can be germinated in a propagating frame within a greenhouse. Many of these devices are now offered for sale at quite low prices, and are a useful investment for any keen gardener. Few alpine seeds will require a temperature higher than 18°C (65°F).

Except where otherwise recommended, all seedlings should be pricked out as soon as they are large enough to handle. Transfer them individually to small pots of the recommended compost, prising their roots carefully from the seed compost and lifting them by their cotyledons (seed leaves). Water each youngster into place, and then stand the pot in a shaded frame for a few days before exposing it to full light.

One final word on seed sowing, or rather two: be patient. Some alpines are very slow to germinate, but if kept gently moist and in the right conditions they will eventually come through. It is not unknown for some seeds to take three years before they emerge. Keep them in the frame and do not worry about severe temperatures – it may be such fluctuation which spurs them into growth in the end.

Cuttings

There are three basic types of cutting: stem cuttings, leaf cuttings and root cuttings. All methods involve a little skill, but are usually accompanied by a fair degree of success even by the beginner.

Stem cuttings are the most common means of propagation and involve the removal of shoot tips, usually in spring and summer. The shoots are usually from 1 to 3in (25mm to 8cm) long, the lower leaves are removed, and a clean basal cut made below a leaf joint before the cutting is inserted in a mixture of equal parts peat and sand (pure sand for some cuttings) either in a garden frame or a propagator. The cutting may be dipped in hormone rooting powder if you feel this will improve rooting (it needs just the slightest dusting on the cut surface).

Some cuttings root more effectively when they are taken with a 'heel' of older wood. Instead of cutting them from the parent plant, pull them away from the main stem so that a piece of harder wood comes with them. This heel can then be trimmed free of any long tail and the cutting inserted in the usual way. Plant your cuttings in pots of a peat/sand mix if you are only raising small numbers of plants, dibbing them in around the edge of the pot so that half the stem is buried. Keep the cuttings in a fairly humid situation while they root by closing the frame light or the lid of the propagator, but keep it shaded from sun or your cuttings will burn. Pots of cuttings can be covered with individual polythene bags, but do check them every day and remove any brown leaves which might result in the entry of fungal infection. Root development is usually accompanied by shoot growth, so when this is observed, tap out and pot up the plants.

Leaf cuttings of plants such as ramondas can be rooted in a similar medium, also in a propagator. Remove a leaf with as much stalk as possible, and insert the entire stalk in the peat/sand compost. Keep the cutting shaded and the atmosphere relatively humid, and eventually a young plant will arise from the base of the leaf blade. Pot it up as soon as it can be conveniently handled.

Root cuttings of plants such as *Verbascum*, *Morisia* and *Primula denticulata* will give rise to new plants. Tap the parent plant from its pot, or dig it up from the soil, and tease away some of the compost to expose the thickest roots. Cut off 1 or 2-in (25mm or 5-cm) sections of root, taking care to keep them the right way up. A flat cut on the upper surface and a sloping cut at the base will indicate top and bottom. (Roots inserted upside down will not grow.) Insert the roots vertically in pots containing a mixture of peat and sand, or in seed trays if you have a good number. The top of each cutting should rest just below the surface of the compost. Keep the cuttings in a shaded frame, and keep the compost gently moist. When shoots appear above the compost the roots can be carefully knocked from the compost and potted up.

Layering

Plants with low or creeping stems will often send out adventitious roots into the soil, eventually forming a thicket of growth. These artificially rooted portions of stem are known as layers, and can be detached and replanted on their own in spring or autumn. You can also help plants along by covering suitable stems with compost or

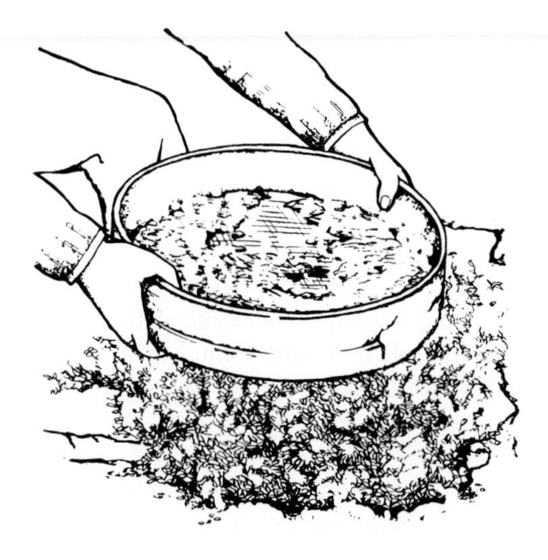

Figure 1.18 Sifting Compost Over a Plant to Encourage Layering

by pegging them to the soil in spring or summer and removing them when they are rooted. Plants which, like the strawberry, form runners make life even easier. At the end of such stems plantlets are produced which root into the soil. It is a simple job to detach such youngsters and replant them to start a new colony.

Division

By far the commonest means of increase, division is also one of the easiest. All clump-forming plants will usually divide easily and re-establish themselves with little difficulty. Spring and autumn are the best times for the job (spring if you are on heavy soil, which tends to hold on to water easily). Lift the plant to be divided and then either cut it into smaller sections with a sharp knife, or prise it apart with your hands. You can use the old technique of pushing two forks into the clump back to back and levering them apart if you like, but I don't know of many alpines that need such heavy treatment, nor of many gardeners who have two forks! Reject any portions of the clump that are dead or exhausted, and replant the selected portions without delay. Each new plant should have plenty of roots and a number of shoots.

Plants such as sempervivums make division easy because they form offsets – little plants that can be cut off and replanted on their own. This type of division can be done at almost any time of year except in the middle of winter.

Of necessity this discourse on propagation has been rather brief, but there are two books which will offer more detailed information if this is needed. *Propagation Of Alpine Plants* by J. K. Hulme is published by the Alpine Garden Society, and *The Complete Book Of Plant Propagation* by Robert Wright and Alan Titchmarsh is published by Ward Lock. I swallow my modesty in recommending the latter.

Chapter 2

A-Z List of Alpine Plants

A

Acaena (Rosaceae) New Zealand Burr

A grand little group of New Zealand natives with ferny, rose-like leaves which are studded with spiky burrs in summer. The brightest species is *A. affinis* (syn. *A. adscendens*), which has a stunning combination of blue-grey leaves and rose-pink young stems. These trail along the ground, but the plant does rise up a little to 6in (15cm) or so. It's a fast coverer. *A. microphylla* is certainly the most ground hugging, its green and bronze leaves making a thick rug of growth that's studded with red burrs seemingly made of spun glass. *A*. 'Blue Haze' is probably the steeliest blue of the collection. Its stems are more woody than most acaenas and are cast along the ground to send up glaucous shoots which eventually form a loose mat. The other species I grow is *A. sanguisorba*, which makes clumps of soft, blue-grey leaves. It's a handsome contrast to darker-leaved plants and brighter flowers.

Cultivation Plenty of sun and a well-drained soil is all these plants ask for. They will grow in the poorest of ground where other alpines have a job to eke out a living and they don't mind chalky soil. Try them in gaps between paving stones on a patio, in alpine lawns and between rocks in beds and on banks. Once established they will quickly romp away, but do remember to give them every chance of sinking their roots into the soil before you neglect them completely – plenty of water will help them to settle into dry spots. Avoid planting acaenas near slow-growing alpines that can be smothered. Winter and spring-flowering bulbs have no difficulty in penetrating the evergreen mats, and so can be carefully planted among the acaenas in autumn.

Propagation Mature plants can be divided in spring or autumn and rooted portions replanted 6 to 9in (15 to 23cm) apart. Seeds can be sown in pots of seed compost in spring and germinated in a garden frame.

Acantholimon (Plumbaginaceae) Prickly Thrift

In spite of the fact that there are around 100 species of *Acantholimon* in the wild, only a handful are cultivated, due mainly to difficulties in propagation. The most well-known prickly thrift, and the most easy to please, is *A. glumaceum*, which makes mounds of narrow, dark-green leaves topped in summer by clusters of pink flowers. It hails from what used to be known as Armenia – now parts of Russia and Turkey. *A. echinus* (syn. *A. androsaceum*) is the most attractive cushion-former of the bunch, making tight hummocks of quite spiny leaves, though it does vary somewhat in appearance. The flowers are again pink. The plant comes from Crete, as does *A. creticum*, spreading itself into Greece as well. It resembles *A. echinus* (of which some folk maintain that it is no more than a form), but makes even tighter hummocks. The last on my list is *A. venustum*, a looser species from Asia Minor which has arching sprays of deep-pink flowers in early summer. This is undoubtedly the most stunning species when in bloom. All have papery calyces behind the petals, which persist when the blooms have faded, and all are evergreen.

Cultivation Only *A. glumaceum* grows reliably well outdoors, enjoying a sunny spot in any well-drained soil, even if it is chalky. If you've a dry, baked spot on your rock garden, try it there. The other three species are best grown in pots in an alpine house or frame. Equal parts of John Innes No. 1 potting compost and sharp sand suits them. Repot in early spring when necessary, but avoid disturbing the roots unless absolutely necessary; all species hate such interference. Water normally from spring to late summer, but very sparingly from September to March.

Propagation Difficult. Seed collected in this country refuses to germinate. Natural-source seed may germinate rapidly if sown in pots plunged in a frame in spring. Cuttings taken with a heel in summer can sometimes be persuaded to root in a heated frame if inserted around the edge of a pot of sandy compost. A more reliable method is to work a mixture of peat and sand among the plant's stems in summer, and hope that some of them will layer themselves. Rooted portions can be severed and potted up the following spring.

Achillea (Compositae) Yarrow, Milfoil

This is a vast genus of plants, many of which are at home in the flower border, but a large number of which performs well on the rock garden. *A. ageratifolia* is a Grecian species up to 6in (15cm) high with toothed, grey-green leaves and white, yellow-centred daisies usually carried singly on the stems. *A.* × *kellereri* is a favourite with many growers, for it produces finely cut grey-green leaves and clusters of white, yellow-centred daisy flowers on 6-in (15-cm) stems. *A.* × *lewisii* 'King Edward' makes a mat of tufted, grey-green leaves and sends up sprays of sulphur-yellow flowers on 4-in (10-cm) stems in summer, and it lasts well. There are many more worth experimenting with; and there are a few not worth looking at.

Cultivation Easy to grow in the right situation, achilleas need a bright, sunny spot in a well-drained but not over rich soil. They are happy in rock gardens, rock beds and among paving stones, but the tufted kinds resent winter wet and a sheet of glass can be used to protect them from excessive sogginess. When happily suited, most achilleas will need to be restrained from swamping their neighbours. Pull off any rampant growths.

Propagation Divide established clumps in spring. Cuttings of soft shoot tips can be rooted in pots plunged in a frame in summer. Pot them up in autumn.

Adiantum (Adiantaceae) Maidenhair Fern

Sneaked in here because of the grace it can impart to the rock garden, *A. pedatum* (North America) is reliably hardy and should make a fine curtain of fresh-green fronds in the right spot. It can make more than 2ft (60cm) of growth in both directions, so give it plenty of room to spread. *A. venustum* (Himalaya), although reputedly tender, can sometimes survive outdoors in a sheltered spot. It is altogether smaller than *A. pedatum* and more welcome for that.

Cultivation Maidenhair ferns need shade and a peaty, leafy soil that is not likely to dry out at any time. Try to give *A. venustum* shelter from the worst winter weather. Both species can also be grown in the earth under the benches of the alpine house, if the soil is suitably prepared. *A. capillus-veneris* may be happy here, too. None of them seem to object to chalky soils, provided there is not a paucity of moisture-retentive organic matter.

Propagation Divide mature clumps in spring. Sow spores on the surface of sterile peaty compost in pans as soon as available. Cover the pan with glass and keep it in a cool, shaded frame, never allowing the compost to dry out. Prick out the youngsters individually as soon as they are large enough to handle.

Adonis (Ranunculaceae) Pheasant's Eye

This romantic-sounding plant is a welcome sight in spring, especially the species *A. amurensis*, which can open as early as February. Its foliage is feathery – like a refined carrot – and its flowers are wide faced and clear yellow. A native of China and Japan, it grows to around 1ft (30cm) high and is most frequently available in its double variety – a curious plant with petals that are often greenish, rather than clear yellow as in the true species. Of the other species, *A. vernalis* is the most easily obtainable. It is European and grows up to 9in (23cm) or so, the bright-yellow blooms opening in March and April.

Cultivation Provided they are grown in a soil which is well drained but never likely to dry out completely, these plants are not difficult. In full sun or dappled shade they seem to perform equally well. Chalky soil is enjoyed, too.

Propagation Divide mature clumps in spring or autumn. Seed can be sown in pots placed in a frame in spring. The fresher the seed, the better. The double form of *A. amurensis* – *A.*a. 'Flore-pleno' – is also reputed to come true from seed.

Aethionema (Cruciferae) Stone Cress

This is an absolute must for any rock garden – especially one on chalky soil. *A.* 'Warley Rose' captivates all who set eyes on it in flower, and it is a worthy reminder of that great, if eccentric, plantswoman Ellen Willmott of Warley Place, Essex, in whose garden it was raised. It was given an Award of Merit by the Royal Horticultural Society as long ago as 1913, and it is still going strong. It makes a spreading shrublet 6in (15cm) high and four times as much across, and is plastered with rich-pink flower clusters from late spring to early summer. In my enthusiasm for this plant I must not forget *A. pulchellum* (Asia Minor to Iran), with its rather glaucous leaves and pink flowers, growing to around 9in (23cm); and *A. grandiflorum* (Lebanon and Iran) which is rather larger, but equally captivating.

Cultivation Plenty of sun and a sharply drained soil will suit aethionemas well. They really enjoy chalky soil, and in very acid ground they will not thrive. Neutral soils can be dressed with lime each spring to keep the plants in good health. All those mentioned can be grown on pots in the alpine house, not because this is a cultural necessity, but simply to add colour to the collection. Grow them in a mixture of equal parts of John Innes No. 1 potting compost and sharp sand, with a little lime added. Water them freely during spring and summer when they are grown in pots, but keep the compost barely moist during autumn and winter. Trim off neatly any wayward stems.

Propagation Seeds of the species can be sown in pots placed in a frame in spring. Shoot tip cuttings 2½in (6cm) long can be easily rooted in June and July. Insert them around the edges of pots filled with equal parts peat and sand, and stand these in a shaded frame. Pot up in late summer.

Ajuga (Labiatae) Bugle

Although it's not everybody's idea of a good rock garden plant, the bugle is useful ground cover in moist and shaded soil at the foot of a rock garden bank where other alpines are proving difficult to grow. *A. genevensis* (6in (15cm)); *A. pyramidalis*, *A. metallica* of gardens (12in (30cm)) and *A. reptans* (4 to 12 in (10 to 30cm)), are all European. The first two are mainly grown for their flowers, which are carried in May in leafy spires. They are usually blue, but may be pink or white in selected forms. *A. reptans* is easily the most invasive, but has some cheery varieties with leaves colourfully marbled. 'Burgundy Glow' is a confection of cream, maroon and pink; 'Rainbow' or 'Multicolor' is purplish-green marbled with orange, and 'Variegata' is grey-green and creamy white. This is the only variety which can be choosy about where it grows.

Cultivation Full sun, dappled or heavy shade are all tolerated, provided the soil never dries out. Boggy conditions are preferred to a sandy desert, and bugles are quite happy on chalky ground. Bright sun produces the best leaf variegations. Over-rampant plants can be torn out to keep the clump in check. Mildew sometimes strikes, but fungicidal sprays will put paid to it. Water the soil well in dry weather, and to get really good cover a spring liquid feed can be given.

Propagation Easily effected by dividing the plants in spring or autumn. Individual rooted plantlets can be transplanted to a spacing of 9in (23cm) to give rapid cover.

Alchemilla (Rosaceae) Lady's Mantle

Hardly noted for their flamboyant brilliance, the alchemillas are nevertheless a pleasant group of garden plants whose main attribute is to hold on to globules of dew as though they were precious diamonds. *A. alpina* is the modest species most frequently offered for rock gardens. It makes a spreading mat 6in (15cm) high, and its leaves are palmate and silvered with hairs on the undersides. The greenish yellow flowers are carried in summer. The plant can occasionally be found growing wild in Britain. Don't break your neck to get hold of it, but give it a chance if it comes your way. It is not such a worthy garden plant as the larger *A. mollis*, which is a must for any herbaceous border.

Cultivation Any ordinary soil and a position in dappled shade will suite the alpine lady's mantle. It will also cascade happily from a rock crevice, but keep an eye on it to make sure that it does not smother more choice alpines. It has a sturdy constitution, but does not enjoy being sun-baked. Chalky soil does not deter it.

Propagation Divide mature plants in spring or autumn.

Alyssum (Cruciferae) Madwort

At the time of writing – early May – a rocky bank in my garden is ablaze with the purple and gold contrasts of *A. saxatile* and aubrieta. It's not the most subtle of combinations, and it's one that any alpine snob would sneer at, but it's a lovely sight! So don't let them turn you against *A. saxatile* just because it's willing to please. Its leaves are spatula-shaped and downy grey – a perfect foil for the acid-yellow flower clusters that are carried up and atop 12-in (30-cm) stems. The plant is a native of Europe, as is *A. montanum*, a more creeping and smaller-leaved species that does not have the upright, clump-forming habit of *A. saxatile*. Its flowers are not such a vibrant yellow, but they are fragrant. If you want your *A. saxatile* diluted, go for the variety 'Citrinum', which is nearer to sulphur yellow. 'Compactum' is smaller in stature, and 'Dudley Neville' has buff-yellow flowers – a bit dirty some would say.

Cultivation You don't have to grow these alyssums – they'll grow you. In any sunny spot in soil that is heavy or light, chalky or otherwise, they perform well. In time the clumps may become untidy and then they are best replaced with youngsters.

Propagation Cuttings of shoot tips 3in (8cm) long can be rooted in a frame in June or July.

Anacyclus (Compositae) Mount Atlas Daisy

A mat-forming daisy with finely cut leaves, *A. depressus* is an easy-to-grow alpine with flowers that are carried singly on 3 or 4-in (8 or 10-cm) stems in May and June. The upper surfaces of the petals are white, while underneath they are rich crimson – the combination of buds and open blooms is pleasantly showy. Out of flower I think the plant is a bit tatty (soggy winter weather rots the leaves), but don't let my ungrateful attitude put you off trying it.

Cultivation A spot in full sun and a sharply drained soil are all that the Mount Atlas Daisy needs. It enjoys life in a scree but resents chalky soil. It can be grown in a pan in the alpine house but it's hardly worth the bother, even though it does enjoy protection from winter wet.

Propagation Seeds can be sown in gritty compost in autumn, and germinated in a frost-free frame. Stem-tip cuttings can be rooted in a frame in early spring.

Anagallis (Primulaceae) Pimpernel

These are pretty plants with blue, red or pink flowers in the species and varieties most commonly available. *A. linifolia* has pimpernel blooms of pure gentian blue, which it carries with great generosity for most of the summer. It is a Mediterranean native which has given rise to several varieties such as 'Collina' and 'Sunrise' – both with red flowers. There is some controversy as to whether or not the variety 'Monellii' is the same as *A. linifolia*, or simply a larger-flowered form. *A. tenella*, the bog pimpernel, is native to Britain, and its form 'Studland', with clear-pink flowers, is well worth acquiring. This species and its form are prostrate, while *A. linifolia* manages to rise a few inches above the ground.

Cultivation *A. tenella* grows naturally in boggy conditions, but it will thrive in ordinary soil which is never allowed to dry out. *A. linifolia* and its varieties do not need such moisture-retentive soil, but they too will resent drought. Many of the varieties are short lived, and may succumb during severe winters. All enjoy a sunny spot.

Propagation Stem cuttings can be rooted in gentle heat in spring and early summer. Seeds can be sown outdoors in spring. It is wise to have one or two young plants ready to replace older ones that may peg out in winter.

Anchusa (Boraginaceae) Bugloss, Alkanet

There's just one species that is coveted by alpinists, and that is *A. caespitosa*. See it in flower and you will instantly see why it is sought after. It makes a neat rosette of long, bristly leaves which is brightened in spring with a posy of stemless, almost gentian-blue flowers. The plant is a native of Crete, and should not be confused with the taller *A. angustissima*, with which it was once officially muddled.

Cultivation Usually grown in the alpine house, anchusa is best in deep pots of a gritty compost composed of one part John Innes No. 1 potting compost to one part sharp sand. It needs repotting every year after flowering, and demands good supplies of water between April and September. Keep the compost much drier in winter. Alternatively, grow it outdoors in a sunny scree bed (limestone or sandstone) and cover the tuft of leaves with a sheet of glass in winter. Scree-grown plants seem to be longer lived than those cultivated in pots.

Propagation This is usually by means of seeds, which are often produced annually. Sow them in pots in autumn or spring, and germinate them in a garden frame. Some authorities state that the plant can also be propagated by short stem cuttings removed in summer, and rooted in a propagating frame with gentle heat.

Andromeda (Ericaceae) Bog Rosemary

Another muddle! All but two species have now been transferred to other genera, and some experts believe that only one species really exists. True enough, most nurserymen offer all their plants as forms of *A. polifolia*, though one – *A. polifolia angustifolia* – we should now apparently refer to as *A. glaucophylla*. All forms have rather narrow, dark, evergreen leaves which are greyish on the undersides, and pitcher-shaped flower clusters of pale pink. The true species grows to about 9in (23cm); *A. p. angustifolia* to about 1ft (30cm) and *A. p. compacta* makes a tighter plant whose stems are less likely to travel horizontally than those of the true species. It has a white-flowered form with greyish leaves which is much prized among alpine buffs. *A. p. nana* is the most popular form, growing to just 6in (15cm). All flower in late spring or early summer. *Andromeda* hails from the colder parts of the Northern Hemisphere.

Cultivation I've grown the plant happily in acid, sandy soil laced with peat, though there is no doubt that it performs best in a peat bed or a peaty part of the rock garden. Shade is preferred to full sun and a moisture-retentive soil suits it best. Cover any sideways-spreading shoots with soil. The plants will not tolerate chalky soils.

Propagation Layer suitable stems in August, and transplant them when rooted in spring. Spreading shoots that have layered themselves can be detached and transplanted in spring or autumn. Seeds can be sown in peaty compost in spring, and germinated in a cool frame. Heel cuttings taken between May and July can be rooted in a mixture of peat and sand in a propagating frame.

Androsace (Primulaceae) Rock Jasmine

The sight of a cushion-forming androsace in flower would convert the most insensitive rugger player into an alpine enthusiast. But I've little room here to ramble romantically about this exquisite group of plants. If you're smitten by their beauty then you cannot afford to be without *Androsaces* by G. F. Smith and D. B. Lowe, published by the Alpine Garden Society. It's a fascinating handbook packed with information on the plants' origins and their cultivation. The plants are divided, botanically, into several sections, but to the gardener there are just two groups: those that can be grown outdoors on the rock garden and those that must be cultivated in pots in an alpine house or frame. All I can do here is to give you my pick of the bunch – the ones that delight the senses every spring, for a few years anyway! Outdoors try *A. sempervivoides*, with its wandering, red-stemmed rosettes and rich-pink flowers; *A. sarmentosa* travels even further with longer stems and has pink flower clusters – each bloom may be almost ½in (13mm) across. Its hairy leaves, though, appreciate protection from winter wet. So, too, do those of *A. jacquemontii*, whose rosettes are exceptionally downy. Its flowers are still rich pink but rather smaller. Of the cushion-forming species suited to alpine house culture I can rave over *A. pubescens* (if you can obtain the true species – I suspect mine of being a hybrid, but it's still good). The cushion is made up of hairy rosettes which are studded with white, yellow-eyed flowers in April or May – the usual time of flowering for most species. *A. pyrenaica* makes the tightest cushions of minute, linear leaves of glaucous blue-grey. Its tiny white flowers have the most delightful jasmine scent. *A. vandellii* (syn. *A. imbricata*) has leaves of silvery grey and white flowers intermediate in size between the two previously mentioned species. Just one more, which is among the easiest of the cushion formers, is *A. cylindrica*. At the moment (early May) my two-year-old plant is a 4-in (10-cm) diameter ball of white flowers radiating on 1-in (25-mm) stalks from the mound of leaves. What a spectacle! The plants are widely distributed in mountainous regions of Europe, Asia and North America. Try any you can lay your hands on.

Cultivation The species that will survive outside insist on a sharply drained soil that quickly disperses excess moisture. They'll enjoy full sun, and the gritty covering on the rock garden will help to prevent their stems from rotting. They thrive in screes and sink gardens. *Most* are tolerant of chalky soil, and *all* will enjoy shelter from wind. All androsaces can be cultivated in pots if preferred, and it's the only way to grow most of the cushion formers successfully. Use a very gritty compost consisting of one part John Innes No. 1 potting compost to two parts sharp sand or fine grit. Dress the surface of the compost with grit or chippings. Remove immediately with tweezers any part of the cushion that dies off, and any dead flowers (unless you can keep a close eye on them for seed production). Repot annually into a slightly larger pot (I always use clay ones for androsaces) immediately after flowering. From March to September the plants can be watered well when they are becoming dry; from October to February water very gingerly, applying a trickle only when the plant begins to look limp. Ventilate well whenever possible, except when heavy rain or fog threatens, to prevent fungus disease attacking. Some experts prefer to water their plants by immersing the pots in a trough of water, and this makes sense. However, I've always watered carefully on to the compost from above (with rainwater), and so far have not had any casualties as a result. Greenfly can be a pest, and should be guarded against by spraying the cushion formers once a month, except when the blooms are open, with a systemic insecticide such as pirimicarb. Benomyl is a suitable fungicidal spray if rotting persists, though look first to your cultivation techniques.

Propagation Seed can be sown in pots of gritty compost as soon as it is available, and germinated in an unheated frame. Single rosettes can be rooted in sharp sand if removed in early summer. Root them in a shaded frame.

Anemone (Ranunculaceae) Windflower

Delightful plants that are easy to grow, anemones should be found a home in any garden. *A. blanda* is my first choice. From its woody rhizomes pale-blue flowers appear over finely divided leaves in February or March. There are also pink and white forms, and a white-eyed magenta form called 'Radar', which I would grow in quantity were it not so prohibitively expensive. The others are cheap and just as garden worthy. The species came originally from Greece and Europe, and another European, *A. apennina*, is also worth having. It blooms in March and April and is similar to our native wood anemone, *A. nemorosa*. This I could never do without, and advise anyone to get hold of its misty-blue variety 'Robinsoniana', which makes large clumps of ferny leaves over which nod the graceful flowers in April and May – rather larger than in the true species. Finally, if you are offered *A.* × *lesseri*, give it a try. It's a robust grower to around 1ft (30cm), with rich-pink or red flowers in May, and is much more beefy than the other three species.

Cultivation *A.* × *lesseri* is happy in any ordinary, well-drained soil and a sunny spot on the rock garden, but give it room to spread. It won't turn up its nose at chalk, but then neither will most anemones. The other three species are best planted in a dappled or gently shaded spot in a peaty, leafy soil or in the peat garden itself. They enjoy a cool, moist root run and prefer to be left undisturbed once planted. The tubers or rhizomes are usually offered in autumn, and should be planted as soon as possible (after an overnight soak if they look at all dry) 3in (8cm) deep and as much apart.

Propagation Lift and divide tuberous and rhizomatous kinds after flowering. Seed can be sown as soon as available (the fresher the better) in pots placed in a cold frame. *A.* × *lesseri* can be divided in spring or autumn.

Antennaria (Compositae) Catsfoot or Cat's Ear

A neat and showy little plant with a mat of short, narrow leaves that are green on the upper surface and white beneath. The species most frequently grown is *A. dioica* which has, in its splendid form *A. d. minima*, 1 or 2-in (25mm or 5-cm)-high downy stems of coconut-ice-pink 'everlasting' flowers carried in clusters of half a dozen or so. Make sure you're not saddled with a form that has white flowers – they're not nearly so cheerful. The plant is a British native in its true form, and may flower between spring and midsummer.

Cultivation Grow *Antennaria* in any ordinary, well-drained or even dry soil and a spot in full sun. It is well suited to making alpine lawns and to sink and patio cultivation. Occasionally it is grown in pots in the alpine house, purely because it enhances a pot-grown collection, not because such protection is necessary. It seems quite happy in my chalky soil in spite of the fact that it is usually found on acid ground.

Propagation Divide mature clumps in spring or autumn.

Anthemis (Compositae) Chamomile

The common chamomile *A. nobilis* is a grand little carpeter for the rock garden, provided the soil below it is not allowed to get dry in the first year of establishment. If you want the plant for its fruity-smelling foliage, go for the flowerless variety 'Treneague', which makes an emerald carpet. The double-flowered variety *A. n.* 'Flore-pleno' deserves to be much more widely grown – its white pom-poms are carried on 3-in (8-cm) stems in summer. Chamomile is from Europe (including Britain), but *A. biebersteinii* (syn. *A. rudolphiana*) comes only from the Caucasus. Its filigree silver foliage is topped in summer with yellow daisies atop 9-in (23-cm) stems. Although *A. cupaniana* grows 1ft (30cm) high and at least 3ft (90cm) across, it's still worth planting for that silvery, feathery foliage and its white marguerite flowers. Like many Italians it has a charm all of its own.

Cultivation All *Anthemis* enjoy full sun and any well-drained soil, but do guard against drought in the first summer or the plants may become very tatty, dying out completely in severe cases. Chamomile makes a good patio plant, and is well suited to creating an aromatic alpine lawn.

Propagation Divide chamomile in spring. Root stem cuttings of the other species in a frame in spring or summer. Seeds, when offered, can be sown direct into a frame in spring.

Anthericum (Liliaceae) St Bernard's Lily

I'll sneak just one *Anthericum* in here, the St Bernard's lily, *A. liliago*. Narrow, grassy leaves 12in (30cm) long accompany a 1½ft (45cm)-long flower stalk of starry, white, yellow-anthered flowers in early summer. It's a lily of consummate grace, and should be found a spot where its size is not a problem. Visit the European Alps and see it in its natural setting, if you've a week or two to spare.

Cultivation This lily has been grown well in a number of different situations – sun or dappled shade; rich and poor soil, but it does demand good drainage, and prefers to be left undisturbed once planted. Among carpeting alpine lawn plants it is shown off to perfection.

Propagation Divide mature plants in spring. Sow seed in a cold frame as soon as it is ripe.

Anthyllis (Leguminosae) Vetch, Ladies' Fingers

These are good plants to bring colour to the rock garden in mid- to late summer. *A. hermanniae* 'Compacta' makes a tight shrublet around 1ft (30cm) tall and rather more across. In summer it is packed so tight with bright-yellow pea flowers that you'll hardly see the foliage. *A. montana* 'Rubra' is rather looser and has rose-red flowers, while the ladies' fingers, *A. vulneraria*, is more herbaceous in habit, with blooms varying in colour from yellow through cream to orange-red. All are European, and the last named is found in Britain.

Cultivation All three demand light, sharply drained soil and a spot in full sun. Chalk is much to their liking. *A. montana* is sometimes grown in the alpine house in pots, in which case a mixture of equal parts John Innes No. 1 potting compost and sharp sand will be suitable. Water freely in spring and summer and much more sparingly in winter, but never let the compost become totally dehydrated.

Propagation Heel cuttings can be rooted in a frame in June, July or August. Seed can be sown in pots in a frame in spring.

Aquilegia (Ranunculaceae) Columbine

These dainty plants are not renowned for their long flowering season, but treasured for the combined effect of their flowers and foliage. There are plenty of dwarfs to choose from, and among the choicest are *A. bertolonii* (Eastern Alps), blue, May; *A. flabellata nana* (Japan), purplish blue and pale yellow, May, and *A. scopulorum* (North America), blue and white, superb glaucous foliage, May to June. None of these three grows much more than 3in (8cm) tall. *A. alpina* is much larger at 12 to 18in (30 to 45cm), and any gardener who buys a plant so labelled takes pot luck. Grown from seed collected in the European Alps, the plants are likely to have lovely blue flowers almost as deep as a mountain sky, but in cultivation, like most aquilegias, this one is no better than it should be – crossing with any other aquilegia in sight. The resulting seedlings are a varied and motley bunch – some good, some rotten. The situation is not helped by nurserymen who label any small aquilegia *A. alpina*. Pay your money and keep your fingers crossed!

Cultivation Grow them all in any gritty, free-draining soil with a helping of leafmould added. Don't disturb them once they are planted. *A. alpina* likes a sun-dappled spot, the others prefer full sun. The three dwarfs are frequently grown in screes or in pans in the alpine house in a mixture of equal parts John Innes No. 1 compost and sharp sand (plus a few limestone chippings or a dusting of chalk for *A. scopulorum*). Water them freely when necessary from March to September; keep them very much on the dry side in winter, giving the slightest trickle when the compost is dusty on the surface. *A. alpina* is especially short-lived, and rather tricky to grow; the others last longer, but are not reliably perennial. Repot in spring only when absolutely necessary. Guard against slug attack – especially in late winter when greedy molluscs can polish off fat, overwintering buds in an evening.

Propagation Sow fresh seed for best results in July or August, in pots that are plunged in a garden frame. *A. scopulorum* is reputedly best sown in October and allowed to be frozen outdoors for a few nights before being placed in a frost-free frame or greenhouse.

Arabis (Cruciferae) Rock Cress, Wall Cress

The common arabis can be seen tumbling in profusion from walls and banks countrywide – a testimony to its reliability and popularity. The commonest form is undoubtedly *A. albida* (syn. *A. caucasica*), so often planted with aubrieta and yellow alyssum. It hails originally from Europe, and as far as Iran, and is now offered in pink and red forms as well as the usual white and its double version. For a gentle change try *A. alpina* (Europe, including Britain), white; *A. ferdinandi-coburgii* (Macedonia), much neater and cushion forming, white. There is also a showy variegated form of the latter, and of *A. albida*. The Californian *A. blepharophylla* is a rosy-red species in its best form (there are deeper and paler strains). Try it and enjoy its daintiness, but remember that it is rather short-lived. All flower in April and May.

Cultivation All do well in full sun in a well-drained rock garden where there is room for them to ramble. Try them also in pavements, patios and dry stone walls, but don't let them loose in sink gardens or they'll swamp everything else. A light clip over with the garden shears after flowering keeps them in trim.

Propagation Divide mature plants in autumn, and replant well-rooted divisions, discarding worn-out sections of the plant. Cuttings with a heel of older stem can be rooted in a frame in July. Seed of

A. blepharophylla can be sown in July in pots plunged in a frame.

Arcterica (Ericaceae) Northern Heath

Northern gardeners, so often at a disadvantage when it comes to the length of their summers, can make southern gardeners wince with envy when they exhibit luscious examples of peat-loving plants that like cool, moist conditions. *A. nana* is one of their treasures. At its best it makes a dense, 2-in (5-cm)-high mound of glossy, evergreen leaves, each one folded quite deeply at the midrib. In April the plant is smothered in creamy-white, bell-shaped flowers just like those of its close relation *Pieris*. Bury your nose among the blooms and savour the sweet perfume. The plant is a native of Japan and that difficult-to-pronounce Russian province Kamchatka.

Cultivation The best plants are grown in a shady spot in a peat bed, where the stems will happily root through peat blocks as well as the peaty soil. If you want to grow the plant in a pan for showing or generally admiring in the alpine house, pot it up in a mixture of equal parts John Innes ericaceous compost, coarse sphagnum peat and sharp, lime-free grit. Keep it plunged in a north-facing frame until it flowers, and then bring it into the greenhouse for its season of glory. Keep it shaded from bright sun, and spray the foliage with rainwater occasionally in summer. Water with rainwater too, never letting the compost dry out, even in winter. Top-dress with the potting mixture in spring, and repot after flowering when necessary.

Propagation Try rooting soft, 1-in (25-mm)-long shoot tip cuttings in a frame in July in a peaty, lime-free soil that is nevertheless quite well drained. Shade the cuttings at all times. Pot up in autumn and overwinter in a north-facing frame.

Arctostaphylos (Ericaceae) Bearberry

The commonest of the bunch is *A. uva-ursi*, which creeps along the ground to make a thick mat of rough, evergreen leaves. Pink-tinged, pitcher-shaped flowers are carried in spring and often followed later in the year by scarlet berries, thereby extending greatly the shrub's season of interest. It's a native of the Northern Hemisphere, including Britain, and well worth growing where it can be given room to ramble.

Cultivation This is another peat lover that needs a lime-free soil It basks happily in full sun or light shade, and puts up with fairly dry soil once it is established. Grow it as part of an alpine lawn, in a peat bed, or where it can sprawl over sheer rock faces.

Propagation Sow seeds in spring in pots of peaty compost plunged in a frame. Take shoot tip cuttings in July and root in a propagating frame with sand and peat in its base; rooting hormone helps. Suitable shoots can be layered in summer and transplanted the following spring.

Arenaria (Caryophyllaceae) Sandwort

I've a soft spot for this unpretentious group of plants, especially for *A. balearica* (Balearic Islands), which makes filmy mats of bright green. Each carpet is dotted with white flowers on 1-in (25-mm) stems from spring right through to autumn. *A. montana* (France, Spain), especially in its variety *A. m.* 'Grandiflora', is rather more showy, throwing up its taller, larger, white flowers in early summer, and *A. tetraquetra* (Spain, Pyrenees) makes tight cushions of peculiarly symmetrical shoots, whose short leaves are angularly arranged opposite one another. Its white flowers open in summer. (See also *Sagina*.)

Cultivation *A. balearica* is a species that is happiest in crevices on the rock garden, or sprawling through chinks in paving where it can be provided with a little shade. It loves a moist soil, and likes nothing more than to be set loose in a murky grotto at the foot of a waterfall – if your landscape is elaborate enough to possess such a feature. The other two species are happy in any well-drained soil and a sunny position. *A. tetraquetra* is often grown as a pan specimen in the alpine house, but I find that unless it is given excellent ventilation and sunlight at all times it tends to quickly lose its tight habit, falling open from the middle.

Propagation Divide mature plants in late spring. Sow seeds in spring in pots of seed compost plunged in a frame.

Arisaema (Araceae)

For a touch of oriental splendour you can't beat the arisaemas. *A. candidissimum* is, I think, the best, for it combines pastel colouring with its bizarre form. A 9-in (23-cm)-high, three-lobed leaf pushes up from the soil in late spring, accompanied by, or slightly later than the flower spike, which is composed of a spathe and spadix, as is usual in this family. Imagine our common cuckoo pint or lords and ladies much more finely formed and intriguingly painted, and you have the essence of the bloom. The spathe is soft pink, veined and striped with white on the inside and flushed purple on the outside. The spadix – the cigar-shaped appendage in the centre – is yellowish green. The plant comes from West China, but seems quite happy in our climate. *A. sikokianum* is Japanese and even more outrageous in appearance. The spathe is purplish brown on the outside, and bends forwards dramatically at the top. The inside is yellowish green, and the spadix is white and swollen at the top so that it resembles a club.

Cultivation *A. candidissimum* is quite hardy and has grown grudgingly for me in sandy earth and a shady spot in Surrey, though for preference it likes a much leafier soil that is not inclined to dry out during summer droughts. It will tolerate full sun if the soil can be kept moist, but it nevertheless demands good drainage. *A. sikokianum* is doubtfully hardy in the north, and repays covering with a sheet of glass in winter wherever it is grown outdoors. It, too, is a lover of leafy soil and shade. Grow the plants in pots if you must, in a mixture of John Innes No. 2 potting compost plus equal parts of well-rotted leafmould and sharp grit, but repot the tubers every year in early spring to give them plenty of mixture to sink their roots into. Keep them cool and shaded at all times, and do not let the compost become bone dry at any time of year.

Arisarum (Araceae) Mouse-tail Plant

This Italian curiosity from the arum family lacks the splendour of *Arisaema*, but will give you more of a chuckle. The plain-green, arrowhead leaves of *A. proboscideum* make a thick, 6-in (15-cm)-deep plantation through which the spathes appear in spring. The flower is just like a thick, brown mouse's rump and tail, and a clump of the plant looks an amusing sight – as if Beatrix Potter's mice had heard Mr McGregor coming down the garden path and taken a dive into the foliage!

Cultivation Grow in any ordinary, well-drained soil which is not likely to dry out in summer. A spot in shade where plenty of leafmould has been worked into the earth is liked most of all. A north-facing bank will also grow the plant well.

Propagation Divide clumps in early spring.

Armeria (Plumbaginaceae) Thrift, Sea Pink

These are indispensable plants for rock gardens, troughs, sinks, patios and even pots in the alpine house. *A. caespitosa* (Spain), makes neat cushions of needle-leaved rosettes from which sprout tufted pink flower heads 2in (5cm) high in spring. 'Bevan's Variety' is a good, deep-coloured form. *A. maritima* is our native sea pink, and is available in many shades from white through pink to rich crimson. The best red I've seen is 'Dusseldorf Pride', which grows up to 4 or 6in (10 or 15cm). *A. plantaginea* is the Jersey thrift, from other parts of Europe as well as from that island. It makes clusters of 6-in (15-cm)-long, ribwort-like leaves and sends up reed-like flower stems about 2ft (60cm) long, each topped with a pom-pom of pink flowers in summer. If you obtain this plant keep it away from a path, where its flowers are kicked off as soon as they are weighty enough to bend the stem down. It's an untidy beast.

Cultivation All love sun and any ordinary, well-drained soil, even over chalk. When grown in pots the cushion formers are happy in a mixture of equal parts John Innes No. 1 potting compost plus an equal amount of sharp grit. Keep them very much on the dry side in winter, but never completely desiccated.

Propagation The cushion formers can be propagated by cuttings rooted in a frame in July and August (each should possess a portion of firm wood); *A. plantaginea* by division in spring.

Artemisia (Compositae)

To most folk the word artemisia is synonymous with silver, and rightly so, for the plant's foliage can be relied on in most gardens to bring lightness to darker plantings. Most are far too beefy to be grown on the rock garden, but two in particular have performed well on my sunny patch. *A. glacialis* (European Alps) is very prostrate and makes a silken mat of shining grey. *A. schmidtiana* 'Nana' is slightly taller, but not much, and just as silvered. It's Japanese and not at all difficult to grow.

Cultivation The plants ask for nothing more than a spot in full sun and any sharply drained soil. Dry earth does not deter them a bit once they are established, but heavy soil might see them off quickly. Winter wet can rot the crowns as well as the foliage, so reduce the dead growth in November by cutting back to leave just a few inches of stem. In really cold districts a pane of glass can be placed over the plants to advantage.

Propagation Cuttings of firm shoot tips can be rooted in a heated propagator in summer.

Arum (Araceae) Italian Arum

Our chalky Hampshire garden is overrun with cuckoo pint, lords and ladies, parson-in-the pulpit, or whatever you want to call *A. maculatum*, but for shady spots on the rock garden where little else will grow, the brighter *A. italicum* 'Marmoratum' (syn. *A. i.* 'Pictum') is well worth squeezing in. The rather crimped, arrowhead leaves are richly marbled with creamy white, and the 1-ft (30-cm)-high greenish arum flowers are followed by clusters of bright-orange berries in autumn. It's a South European plant that offers a long season of interest – particularly to flower arrangers.

Cultivation Try to buy a plant that is already in leaf – that way you can make sure it is what it purports to be. I've made the mistake of buying a dormant plant in the past, only to discover that its leaves were poorly marked. Heavy, moisture-retentive soil (not at all like that found in any good rock garden!) is most to their liking, and they will put up with full sun, but prefer at least dappled shade and at most deep gloom. Good in dull, north-facing spots.

Propagation Divide established clumps in autumn.

Asarina (Scrophulariaceae)

I first saw this useful plant scrambling through the base of a hedge at Wisley, where somehow it managed to eke out a living among the dry, root-infested soil. It's a treasure for that parched and difficult spot in the rock garden where most plants are impossible to establish. *A. procumbens* (syn. *Antirrhinum asarina*) is its full name. It comes from the South of France, grows but a few inches high, and has rounded, hairy leaves and snapdragon flowers of pale yellow marked with red. It will bloom freely in spring and summer, and amply earns its keep, scrambling over the ground quite quickly.

Cultivation Plant *Asarina* at the foot of shrubs on the perimeter of the rock garden, in a dry wall or the side of a raised bed. It enjoys a spartan diet, but should be watered well until its roots are established enough to search for their own nutrition. It prefers shade to full sun.

Propagation Sow seeds in pots placed in a garden frame in spring. Root stem cuttings in a frame in June.

Asperula (Rubiaceae) Woodruff

Although our native sweet woodruff, *A. odorata*, is a woodland shade lover, the other species in the genus tend to prefer full sun. Use the native for shady spots where it will produce a 9-in (23-cm)-high thicket of leaf rosettes which are sweetly scented – especially when dried and crushed. White, fragrant flowers appear at the stem tips in early summer. Quite a different kettle of fish is *A. suberosa* (sometimes listed as being synonymous with *A. arcadiensis*, which is, in fact, rarer and rather more difficult to grow). *A. suberosa* is a Greek native and possesses small, downy, grey-green leaves held on 3-in (8-cm) stems in the form of a mound (which will fall apart if the plant is not kept happy). Starting in early summer the plant may be smothered in tiny, long-tubed, four-petalled flowers of soft pink, carried in clusters at the stem tips. A well-grown plant in full flower is captivating.

Cultivation Grow sweet woodruff in a shady spot below trees or shrubs. It romps away with no trouble at all in any soil. *A. suberosa* is usually grown in pans in the alpine house or frame, but it can succeed outdoors in a sunny scree or sink garden or even in the wall of a raised bed. In pots give it a mixture of sharp grit and John Innes No. 1 potting compost in equal parts. Water freely when necessary from April to September, taking care not to saturate the hairy leaves, but keep the compost *almost* bone dry in winter. Pluck up the courage to shorten the stems of the plant by about half after flowering – that way you'll keep a tight mound of foliage. The plant likes a chalky soil, and limestone chippings can be added to the potting mix or to the soil in which it is to grow in a scree or trough. A pane of glass will afford these outdoor plants welcome protection in winter, but it is not essential in sheltered spots. Repot container-grown plants in spring when necessary, but don't over-pot them – they can't stand a mass of cold, wet compost.

Propagation Remove shoots arising from below ground level in spring and early summer, and root in a frame in pure sand. *A. odorata* can be divided in spring or autumn.

Asplenium (Aspleniaceae) Spleenwort

These useful ferns bring fresh greenery to rock gardens, walls and screes. *A. scolopendrium* (syn. *Phyllitis scolopendrium*) is the hart's tongue, and its bold, bright-green, strap-shaped fronds look a treat when unfurling in spring. They are evergreen, but are best chopped off completely when they become old and battered. *A. s.* 'Crispum' is a fine form with corrugated fronds. Both plants grow about 1ft (30cm) high, much taller than the trim *A. ruta-muraria*, the wall rue spleenwort, which makes only about 3in (8cm). Treasure it in a wall. *A. trichomanes*, the maidenhair spleenwort, and *A. adiantum nigrum*, the black spleenwort, both have feathery fronds more or less than a foot (30cm) long. All are British natives.

Cultivation All three enjoy well-drained soil, and are most easily established when grown from young plants (though *A. scolopendrium* can be moved even when it has made a hefty crown). All prefer gentle shade to full sun. *A. ruta-muraria* grows best in a wall or a limestone scree, and *A. trichomanes* enjoys limestone chippings mixed into its growing medium, too.

Propagation Divide mature plants carefully in spring and replant divisions immediately, spraying with rainwater to aid establishment. Spores can be sown in pots placed in an unheated frame exactly as for *Adiantum*.

Aster (Compositae)

There's a fair collection of asters suitable for rock-garden culture, but by far the most popular is *A. alpinus*, a European alpine most commonly offered in the variety 'Beechwood', which possesses lavender-blue daisy flowers centred with a golden boss. It grows between 4 and 6in (10 and 15cm) high, as does *A. natalensis*. This South African is a purer blue and has a more orange centre. Both species make spreading clumps and flower from June onwards.

Cultivation Asters love sun, and they need a well-drained soil if they are to thrive. Give them a bright spot on the rock garden where they will have room to spread a little without becoming a nuisance. Chalky soils are easily tolerated.

Propagation Divide mature clumps in spring.

Astilbe (Saxifragaceae) Spiraea

More common in the bog garden than in the rock garden, astilbe nevertheless has some superb dwarf forms to cheer the alpine gardener. For summer colour try *A. chinensis* 'Pumila', a Chinese native with rich, rosy-red spires of bloom on 9-in (23-cm) stems. *A. glaberrima* (Japan) is most often sold in the form known as 'Saxosa'. It's just 6in (15cm) high, and has fluffy flowers of a pleasing pink. There are plenty of hybrid astilbes, but have a go with 'William Buchanan', whose leaves are tinged wine red, and whose flowers are creamy white. 'Sprite' is another cracker that's pink and no more than 9in or 1ft (23 or 30cm) high. All astilbes have delightful ferny foliage.

Cultivation Although the rock-garden astilbes don't demand boggy conditions, they do prefer a spot where they are not likely to be sun baked. The soil should be well drained, but not likely to dry out excessively at the height of summer. Dappled shade seems to suit their constitution better than full sun. They don't mind chalky soil.

Propagation Divide mature clumps in spring.

Aubrieta (Cruciferae) Purple Rock Cress

Who'd be without a bank of bright aubrieta in spring – not me! Don't let plant snobbery put you off this glorious carpeter, which rewards its cultivator annually with a most generous show. The plants grown are, in the main, hybrids of *A. deltoidea*, from Southern Europe. There are hundreds of them in shades of lavender, purple, mauve, crimson and pink, and one or two with white blooms. The leaves are slightly downy, and make thick mats perfectly designed to show off the crowds of four-petalled flowers. Take your pick of the varieties from any alpine catalogue – few of them will disappoint.

Cultivation Superb in vertical walls or rock-garden banks where they can travel downwards. Chalky soil is enjoyed, and sun is essential, but the plants are happy in any reasonable spot in any part of the country. Clip them over after flowering to avoid that ugly, stringy appearance common in neglected rockeries.

Propagation Divide mature clumps in spring. Cuttings of shoot tips can be rooted in a propagator in late spring and summer.

Blechnum (Blechnaceae) Hard Fern

The British native hard fern is *B. spicant* – at around 1½ft (45cm) it may be considered too gross for all but the largest of rock gardens, but it is nevertheless graceful with its deeply serrated fronds. Much more accommodating, and more easily accommodated, is the delightful *B. penna-marina* from New Zealand. It hugs the ground and sends up thick ranks of saw-edged leaves just 4 or 6in (10 or 15cm) high. Its dark-green mat is a fine foil for brighter flowers.

Cultivation *B. spicant* must have lime-free soil, but *B. penna-marina* is not so fussy. However, both do enjoy a shady spot in a cool, leafy soil that is not likely to dry out. Plant them to bridge the gap between shrubs and the rock garden in patches of earth that most rock plants would find too gloomy. Cut off brown fronds in late winter before the new crop emerges.

Propagation Divide mature clumps in spring. Sow spores in pots as for *Adiantum*.

Bletilla (Orchidaceae)

Orchids are not to everyone's taste, but they do provide a talking point when grown outdoors. *B. striata* (syn. *Bletia hyacinthina*) is not too difficult outdoors in the right spot, and it should delight you with its show of lilac-purple flowers in summer. It's a Chinese species, and grows to 1ft (30cm) or more in height. The sword-shaped leaves are fresh green.

Cultivation The most reliable outdoor clumps are cultivated at the foot of a warm, south-facing wall or bank in soil that has been leavened with grit, leafmould and peat so that it is open and well drained, yet not likely to dry out. A few panes of glass will afford welcome additional protection in winter. The plant can also be grown in deep pots or pans in an alpine house. The compost should be a mixture of equal parts John Innes No. 1 potting compost, sieved leafmould and sharp grit. Water the compost freely from spring to autumn, and keep it just slightly moist (not dry) through the winter. The plant is not totally hardy, and is certainly more reliably overwintered in a garden frame or alpine house. Repot in early spring when necessary.

Propagation Divide mature clumps after flowering.

Bruckenthalia (Ericaceae) Spike Heath

There's only one species of *Bruckenthalia*, and that is *B. spiculifolia*; but it's well worth growing for its summer show of pink bell flowers. The foliage is heathery, and its June-July flowering season makes it popular with heather fanciers who are on a lean diet of flowers at that time of year. The plant is an alpine from south-east Europe and Asia Minor, but it is very much at home in Britain provided the soil is to its liking. It's a good spreader and makes thick mats up to 6in (15cm) high.

Cultivation Lime-free soil is essential, and a peat bed produces the best plants. *Bruckenthalia* enjoys best a spot in full sun, provided the soil in which it grows is not likely to dry out. Top-dress older plants with a peaty, leafy soil so that they will layer their shoots and make division easier.

Propagation Divide mature plants in spring, replanting rooted portions at a spacing of 1ft (30cm). (Layering is best carried out in summer, and the plants should be ready for dividing the following spring.) Cuttings of soft shoot tips can be rooted in a propagating frame in summer. Seeds can be sown in pots of sandy compost in spring, and germinated in a cool frame.

Calandrinia (Portulacaceae) Rock Purslane

Few flowers are more brilliant than those of *C. umbellata*, which opens its magenta blooms in succession through the summer. It's a native of Peru, and in spite of its truly perennial nature it is often treated as an annual, thanks to the British weather, which tends to shorten its life. But it's still worth growing where a bold bunch of flowers is wanted. It seldom grows more than 6in (15cm) high, but its trailing stems and narrow, succulent leaves will cover the ground effectively.

Cultivation It undoubtedly grows best in a scree, where the sharp drainage will be to its liking. Full sun is essential and chalky soil does not worry it. Plant it where its stems can cascade over a retaining wall or a sheer rock face. It can also be grown in pots of gritty compost in the alpine house, but it will not necessarily be longer lived. Water freely when dry in spring and summer but keep very much on the dry side in winter.

Propagation New plants can be raised from seeds in February or March. Sow them in pots of compost placed in a warm propagator. Plant out the youngsters when the danger of frost is past. A pane of glass will ward off excessive winter wet and help the plants to last a second year.

Calceolaria (Scrophulariaceae) Slipper Flower

If you consider the florist's calceolaria to be vulgar, don't be put off growing the alpine types, for they are in a different class altogether. The most cherished is undoubtedly *C. darwinii*, whose open-fronted pouches are chrome yellow, spotted with mahogany, and possessing a central bar of white. All this on a stem no more than 4in (10cm) high. 'Walter Shrimpton' is a hybrid derived from *C. darwinii* and *C. fothergillii*, and it appears to be more robust than either of its parents. It has even more mahogany and rich orange in its make-up. *C. darwinii* hails from the Magellan Straits, and the less strikingly coloured *C. fothergillii* from Patagonia and the Falkland Islands. *C. falklandica*, whose origins are obvious, is not so showy as the others mentioned here, but its daintiness commends it, and it is very much at home in one of my sink gardens in John Innes compost plus grit. It has 9-in (23-cm)-high stems, atop which are carried from two to four pouches of bright yellow, spotted with mahogany only in the throat. It has rounded, broader leaves than the other species mentioned. All flower from June onwards.

Cultivation All these calceolarias do best in fairly moisture-retentive soil that is not likely to dry out in summer. Work in plenty of grit, but also add generous amounts of peat and leafmould to keep the roots cool. Chalky soil is not much enjoyed. They all like good light but not scorching sun. Try them in part-shaded ledges on the rock garden or in sinks and troughs. Watch out for greenfly, which nestle among the leaves, and for red spider mites, which cause the foliage to become papery and bleached.

Propagation Divide large clumps in spring. Sow seeds in gentle heat in spring, in pots of seed compost placed in a propagator. The seed is dust fine; sow it on the surface of the compost and leave it uncovered, except for a sheet of glass and paper.

Callianthemum (Ranunculaceae)

This is a charming little plant; some growers find it difficult to flower regularly, others succeed annually. There are two species common in cultivation: *C. rutifolium* (syn. *C. anemonoides*), and *C. kernerianum*, both of which are from the European Alps. The first carries pure-white daisies, flushed with pink on the undersides, over its aquilegia-like leaves, and the second is not so full petalled and rather more lilac in colour. Both have a central boss of yellow florets, and both grow between 4 and 6in (10 and 15cm) high. *C. kernerianum* flowers in May, and *C. rutifolium* in April.

Cultivation In the open the plants flourish best on a scree where drainage is sharp and the light intensity high, but they are also grown in pots in the alpine house. John Innes No. 1 potting compost and sharp grit mixed together in equal parts suits them fine. They'll need plenty of water in the growing season – from March to late summer – but keep them barely moist in winter, though not bone dry when the foliage has died off. Repot every second year after flowering.

Propagation Mature plants can be carefully divided in June.

Campanula (Campanulaceae) Bellflower

Open any alpine plant book to look up *Campanula* and you'll find page after page of different species, most of them reliable beauties suited to cultivation in almost any situation, and offering a long show of dainty, blue or white bells. I'm not going to ramble on at length about all these campanological treasures; other writers have done the job more comprehensively than I could hope to. But I will mention a few bellflowers I think you should not be without. For sheer ease of culture combined with a generous floral display, *C. carpatica* and its forms takes some beating. Originally from the Carpathians, it has deep and wide-faced bells which are carried, upturned, on stems 6 to 12in (15 to 30cm) high. There are lavender-blue and white varieties, and 'Chewton Joy' (pale blue) is a deservedly popular form. The two campanulas with virtually unpronounceable names – *C. portenschlagiana* (syn. *C. muralis*) and *C. poscharskyana* – grow in spite of the gardener rather than because of him. Both form thick, spreading mats of greenery covered with blue bells in summer. The first comes from Dalmatia and the second from Eastern Europe. *C.* X *haylodgensis* is an interesting little hybrid which forms not bells, but little, pale-blue rosettes of double blooms. It's a double-flowered plant which still manages to retain a vestige of gracefulness. There are other choice species, some of which demand alpine-house culture, and some of which are short lived, but this is usually because they are so generous with their blooms that they flower themselves to death.

Cultivation In general all the campanulas suited to outdoor cultivation like an open, sunny spot and any reasonable soil that is well drained. Most of them tolerate or even enjoy lime, but *C. cenisia*, *C. excisa*, *C. hawkinsiana*, *C.* X *haylodgensis*, *C. piperi* and *C. sartori* prefer lime-free soil. Grow the trailing kinds where they can cascade over rocks and dry walls, and the clump-forming kinds in crevices among rocks. Watch out for slugs, which can decimate young plants in particular. Choice campanulas can be grown in the alpine house in pots containing a mixture of equal parts sharp grit and John Innes No. 1 potting compost, plus a light sprinkling of lime or a few limestone chips if they are lime lovers. Water freely in summer when growth is underway, but be careful in winter, keeping the soil only slightly moist. Slug attacks are common in winter too, so take precautionary measures in the alpine frame.

Repot container-grown campanulas every spring just as growth commences.

Propagation The clump-formers and spreaders are easy to divide and re-establish in early spring. Other species can be raised from seeds sown in pots of seed compost in spring, and germinated in a garden frame. Like primulas, campanulas germinate best if they are exposed to light, so do not cover the seeds – simply press them into the surface of the compost. Some species will root readily when propagated from summer cuttings rooted with gentle heat in a propagator; others refuse. Trial and error will reveal the possibilities.

Carduncellus (Compositae)

This is a curiosity worth growing both for its flowers and its leaves, though it can be absolutely ruined by slugs in damp weather. *C. rhaponticoides* (Morocco) makes a symmetrical rosette of dark-green, purple-tinged leaves which lies flat against the soil, and from which arises in early summer a stemless head of fluffy, lavender-blue flowers – rather like the top of a thistle. This 'shaving brush' stays for a few weeks then fades, leaving the rosette to multiply during the rest of the year. It's hardly the showiest alpine, but it's modest in its requirements and takes up little space.

Cultivation Any ordinary soil that is reasonably drained will suit *Carduncellus*, and a spot in full sun is essential. It tolerates a good baking, and is worth trying in arid parts of the rock garden.

Propagation Divide large clumps into individual rosettes in early spring.

Carlina (Compositae) Carline Thistle

Some gardeners spend most of their time discouraging thistles, but here's one well worth encouraging. *C. acaulis* (Europe) won't grow to 4ft (1m 25cm) like the common thistle; instead it forms a spiny rosette of dark leaves pressed flat against the ground. The showy, single flower is stemless, or on a very short stalk, and is creamy white in colour. It appears in July or August.

Cultivation Be cruel to it for best results. Plant it in a dry, sunny spot in a scree bed or poverty-stricken earth in the rock garden. Rich, moisture-retentive soil will produce an overfed, untypical plant that is reluctant to flower. It's a good alpine lawn plant, too, but don't walk on it with bare feet.

Propagation Seeds can be collected in late summer, and sown the following spring in pots of compost placed in a garden frame. It makes sense to prevent youngsters from flowering in their first year, for if they possess but a single rosette this may die after flowering. The plant will perish if no smaller rosettes are ready to take over.

Cassiope (Ericaceae)

Nobody could call the cassiope flamboyant, but a well-grown plant in full flower is a beautiful sight. In all forms the scaly leaves are tight pressed against the stems, and white, bell flowers, like those of pieris, are carried (with any luck) in April or May. One or two new hybrids seem to arise every year, but the species still retain their popularity. *C. lycopodioides* (north-west America and north-east Asia) is very prostrate and mat-forming and is especially fine in its form 'Beatrice Lilley'. *C. wardii* (Tibet) is the tallest species at up to 1ft (30cm), and its leaves possess white hairs which may impart an overall greyish tinge. Of the hybrids, *C.* X 'Muirhead' is the most popular, and makes a plant around 6in (15cm) high, while 'Kathleen Dryden' warrants inclusion in every collector's set for her horizontal, spreading habit.

Cultivation The cassiopes are peat lovers for a deep, moist, acid soil. They tend to flower best in a bright spot where there is nevertheless plenty of moisture available for their roots. A peat bed which faces north, but which is not overshadowed by trees, will produce good plants. The carpeting types can be top-dressed with a mixture of peat and sand in late May to keep them in good condition. Cassiopes are often grown in pans for showing, but these must be plunged outdoors in a shaded frame all the year round, except when in flower when they can be brought into an alpine house to admire. Pot up the plants in a mixture of equal parts sieved leafmould, John Innes ericaceous mix and coarse grit. Water with rainwater, keeping the compost gently moist at all times, even in winter. A daily spray with rainwater is enjoyed in pot-grown plants in spring and summer. Repot after flowering when the plant has outgrown its container; until then content yourself with applying a peat and sand top-dressing each spring as growth commences.

Propagation Seeds can be sown on the surface of a compost consisting of equal parts peat and sand in early spring. Germinate them in a shaded frame, taking care to keep the compost moist. Cuttings are not the easiest things to strike, but if they are taken from young shoots with a heel of older wood in summer, success is possible.

Root them in peat and sand in a shaded frame, and spray them daily with rainwater. Rooted portions of the carpeting types can be removed and transplanted in spring.

Celmisia (Compositae) New Zealand Daisy

Increasingly popular Australasian plants with handsome foliage, the celmisias provide a pleasant change on the rock garden, and a few of them enjoy life in pots. All the following are from New Zealand, and represent some of the best kinds in cultivation: *C. coriacea*, 9in (23cm), leaves silvery above and thickly felted beneath, white daisy flowers carried in June; *C. argentea*, 3in (8cm), tufts of silvery leaves, stemless white daisies at the shoot tips in May; *C. bellidioides*, 3in (8cm), dark-green mats of leaves and white daisy flowers in spring and summer; *C. spectabilis*, 1ft (30cm) (flowers to 1½ft (45cm)), bold, upright rosettes of green leaves which are silver haired above and creamy felted beneath; its white daisy flowers are very large.

Cultivation You'd think from their appearance that these plants would need a poor, dry, well-drained soil and scorching sun, but strangely, they seem to relish a soil that is on the rich side and not likely to dry out. The sun they do enjoy, and a sheltered position they insist on if they are to come unscathed through British winters. It's worth protecting each plant with a sheet of glass to keep off winter rains, which can rot the crowns. *C. argentea* and *C. coriacea* can be grown in pots in the alpine house in a mixture of equal parts John Innes No. 1 potting compost and sharp grit. Repot them every year after flowering, and give them plenty of water during spring and summer, much less in winter to prevent rotting.

Propagation Seeds can be sown in pots of seed compost in spring, and germinated in a frame. Cuttings of side rosettes can be rooted in an unheated propagator in summer.

Celsia (Scrophulariaceae)

There's just one species of *Celsia* suitable for cultivating on the rock garden or in pots. *C. acaulis* is a Greek treasure which makes a flat rosette of rough, deeply veined leaves, in the centre of which appears a cluster of yellow mullein flowers at any time between spring and late summer. The flower stems are seldom more than 3in (5cm) high, making the plant a well-proportioned beauty that no alpine enthusiast should want to be without.

Cultivation Best in a sunny scree or a patch of sharply drained earth on the rock garden. In pots the plant enjoys a mixture of John Innes No. 1 potting compost and sharp sand. Water freely when dry in spring and summer, but be more careful in winter, keeping the compost barely damp. Repot annually in spring.

Propagation Divide plants in early autumn, cutting up the crown if necessary into sections – each with a bud and some roots. Seeds can be sown in pots placed in a frame in spring, but the progeny are likely to be variable, for the plant hybridises readily with its close relation *Verbascum*.

Cerastium (Caryophyllaceae) Snow in Summer

'A thug and a strangler' – so speaks Will Ingwersen of *C. tomentosum*, that ubiquitous white-flowered, white-leaved carpeter of many a suburban verge. 'This unpromising race of weeds', says Farrer; while Clarence Elliott confesses 'It took me half a lifetime to find out that I do not really care for Cerastiums'. It's a wonder anyone grows these obliging carpeters. True enough, many of them are not fit to let loose among your alpine treasures, but where they can be given a bit of space to roam and riot they look a delight in spring. All species are carpeting to a greater or lesser degree, and all have white flowers.

Cultivation I'm tempted to say 'any soil, any site, anywhere', but feel impelled to be fair to the plant and recommend that it be grown in full sun in a reasonable soil. A clipping over after flowering will prevent that straggly habit common in older plants. Whatever else you do, keep cerastiums out of sink gardens. They don't mind chalk one bit. A final word from Clarence Elliott: 'I conveyed my last plants, many years ago, to a chalk-cutting on the railway line between Stevenage and King's Cross, where they have ample scope and are a great success.'

Propagation Divide clumps in spring, or, in the words of Lawrence Hills, 'Tear them up by the roots and plant them where needed, or burn by the barrow-load.' Why are gardeners so scathing of plants that enjoy life?

Ceratostigma (Plumbaginaceae)

Ceratostigma plumbaginoides is a Chinese herbaceous plant, valued on the rock garden for its show of gentian-blue flowers in late summer and early autumn – often a thin time for flowers. It obliges further by turning its foliage bright shades of red in autumn, and it spreads by means of underground stems – growing anything up to 1ft (30cm) high.

Cultivation In a sheltered spot in sun or gentle shade the plant will thrive, provided the soil is well drained. Cut away any dead stems in spring. Wherever you plant it, give it some room to spread.

Propagation Divide in spring or late summer. Cuttings can be rooted in summer in a frame containing peat and sand. Take young shoots with a heel of older wood.

Cichorium (Compositae) Chicory

I'm not advocating that you should grow chicory on the rock garden, rather that you should grow one of its curious relations in the alpine house or frame. *C. spinosum*, from Spain, Greece and Sicily, makes a thick hedgehog of spine-like stems. In summer masses of bright-blue flowers appear to briefly turn the inhospitable dome into a mound of flower.

Cultivation This chicory has been grown outdoors, but it seldom lasts for long. If you want to chance it, grow it in extremely well-drained soil in full sun. It will enjoy scree conditions and happily puts up with chalk. In the alpine house or frame it can usually be guaranteed a longer life, though severe winters of the 1981/2 calibre may well kill it. Give it a compost composed of two parts sharp grit to one part John Innes No. 2 potting compost, and water it carefully at all times, letting it dry out a little between visits with the can. Excess moisture can be fatal at any time of year, but especially in winter. Cut out any obviously dead stems in spring.

Propagation Cuttings of young shoots can be rooted in a propagator in May and June. Root cuttings can also be taken in spring, and started into growth in boxes of peat and sand placed in a frame.

Codonopsis (Campanulaceae)

The best place for these scrambling or twining plants is among rock-garden shrubs, where they can respectably wander without ever threatening to take over. *C. clematidea* (Asia) has whitish bells suffused with pale blue and it scrambles to a height of 4ft (1m 25cm) or so. Look up into its bells and you'll see darker markings. *C. convolvulacea* (China and Himalaya) is a better blue and slightly tender, scrambling to around 3ft (90cm). It is often grown as an alpine house plant and trained up a support of twigs. *C. meleagris* (China, Yunnan) has greenish-yellow bells veined with rich brown. Stick your nose among the flowers of most species and you'll be far from impressed; the fragrance is less than admirable. All flower in summer.

Cultivation A sunny bank or even a scree bed will suit these plants, for they adore sharp drainage. When grown in pots the plants enjoy the usual mix of equal parts John Innes No. 1 potting compost and sharp grit. Water them fairly freely from spring to autumn, but then withhold water as they die down, applying just a trickle to the compost from time to time to prevent total dehydration. Cut off the dead growth an inch or two (2 or 5cm) above compost level. Outdoor plants should be clearly marked with a label during winter so that they are not disturbed during cultivation or planting operations.

Propagation Sow seeds in pots of seed compost in March, germinating them in a frame which can be kept frost free. You can take a chance and prick out the seedlings as soon as they are large enough to handle or you can leave the young plants in the pot until they die down in autumn and then knock out the compost, search for the small but swollen roots and pot these up individually with no delay. Desiccated roots quickly die.

Convolvulus (Convolvulaceae) Bindweed, Devil's Guts

Although the common bindweed is a close relation of the following cultivated kinds, they should not be neglected, for they show nothing of its insidious and rampant habits. *C. cneorum* (southern Europe) is not strictly a rock-garden plant, and is too large for all but the mammoth rock bank or bed. But it is so lovely that I cannot leave it out. Its leaves are thickly coated with silvery silky hairs, and the white trumpets, banded pink on the outside, open at the stem tips in summer. In time, and with kind winters, it will make a shrub 3ft (90cm) across and 18in (45cm) high. *C. mauritanicus* (syn. *C. sabatius*) (North Africa) is more easily accommodated. It's a low- and fast-growing spreader with oval, green leaves and eye-catching, dusty blue trumpets that blow their fanfare from summer to autumn. *C. nitidus* (syn. *C. boissieri*) (Spain) is the trickiest to grow and the most difficult of the three to find. The leaves make a thick mat around 2in (5cm) high, and are covered with silky white hairs. The flowers (if you can persuade the plant to produce them) appear in summer in modest quantity. They are a delicate shade of pink.

Cultivation All these convolvuluses like sun and shelter, and in severe winters they may be killed outright. Don't let that put you off giving them a try. Plant *C. cneorum* where it has room to spread a little over the years. Any well-drained soil and a spot in full sun will suit it and it doesn't mind chalk. As it ages it will be impossible to cover it with the usual pane of glass in winter, but a few rooted cuttings will provide you with insurance if they are kept in a frame through the winter. Cut out any dead growths in spring. Find *C. mauritanicus* a bright spot on a scree bed or a sunny patch on the rock garden. It spreads madly in my chalky soil. *C. nitidus* can be risked on a scree bed, but it really needs alpine-house treatment. Pot it in equal parts John Innes No. 1 potting compost and sharp grit, and water it freely when dry from spring to autumn, keeping the compost very much on the dry side in winter. Repot every other year in spring.

Propagation Cuttings of shoot tips can be rooted in a propagator in summer. Seeds can be sown in pots of seed compost in spring and germinated in a garden frame.

Coprosma (Rubiaceae)

You're likely to come across only one member of this genus suitable for rock gardens, and that is *C. petriei*, a dense mat-former from New Zealand. It grows just 3in (8cm) high, but may spread sideways for up to 3ft (90cm). The leaves are small, oval and dark green, but it is the translucent fruits that are the main attraction. They may be anything from white to purple and appear in late summer. Male and female flowers appear on different plants, so the fruits are carried only by females. Gardeners who think that the berries are horrible (because they look like massive slugs' eggs) would be better off with a male, but then there's only the thyme-like foliage to admire, for the flowers are quite insignificant.

Cultivation Find *Coprosma* a sunny ledge or crevice on the rock garden, or make it part of an alpine lawn. It grows well in a variety of soils, but enjoys especially lighter ground that has been enriched with leafmould. Once planted it prefers to be left undisturbed for years. The plant may be killed in very severe winters.

Propagation Heel cuttings can be rooted in a propagating frame in summer.

Cornus (Cornaceae) Cornel, Dogwood

The creeping dogwood, *C. canadensis* (or if you wish to be really up to date, *Chamaepericlymenum canadense*) is a North American plant well suited to being grown in a peat bed. It makes 6-in (15-cm)-high ground cover of oval green leaves that are deeply veined. In May or June the small flowers open, each one surrounded by four, smart, white bracts. If you're lucky some red berries might follow in autumn, but even if you're not, the flowers and leaves alone are a pleasing show.

Cultivation This is a good-looking plant that's easy to grow in deep, moist, peaty soil in sun or gentle shade. Enrich your soil before planting with peat and leafmould – the plant is native to boggy areas and resents drought. It dislikes lime, too. Give it room to spread.

Propagation Divide clumps in autumn and replant portions at a spacing of 1 to 1½ft (30 to 45cm) if ground cover is wanted.

Cortusa (Primulaceae)

Closely related to the primulas, *C. matthioli* is native to a wide area – from the western Alps as far east as Siberia. Its leaves are deeply lobed and hairy – not unlike those of heuchera – and the nodding, bell-shaped primula flowers are carried in clusters on 8-in (20-cm) stalks over the mound of leaves. The blooms vary in colour from rosy pink to purple, and they usually open in July.

Cultivation *Cortusa* is most at home in a peat bed in gentle shade, where it can revel in the cool, moist conditions it loves.

Propagation Divide clumps in spring, replanting the divisions without delay.

Corydalis (Fumariaceae) Fumitory

The corydalises are a most beautiful group of plants, but the gardener could be forgiven for thinking of them as either very tricky, or so rampant that they soon become weeds, for it is the species at these two extremes that are remembered. I find myself drastically pruning my list of recommended species to fit available space, but the following would be my 'Desert Island Corydalises', even though they number less than eight! Top of most people's list would come *C. cashmiriana* (Himalaya) with its finely cut grey-green leaves and clusters of sky-blue flowers (if you're lucky) in April or May. It's a tricky one but an irresistible challenge. So too is *C. transsilvanica* (origin obscure), which is about the same height at around 6in (15cm). Here the flowers are a rich rose pink and very striking. *C. solida* (northern Europe and Asia) is easier altogether, and bears plenty of purplish flowers on 6-in (15-cm) stems in Spring. *C. cheilanthifolia* (China) is worth growing for its foliage alone – it's as feathery as any fern and you could easily mistake it for one if its yellow spring and early summer flowers were not in evidence. It grows to about 9in (23cm) high. It seeds itself around with reasonable freedom, but nothing like so perniciously as *C. lutea* (Europe, including Britain), which is superb in walls and pavements, but not where choice alpines are being grown. Introduce it at your peril!

Cultivation All corydalises are worth trying in a peat bed, where they enjoy the gentle shade and cool root run offered. *C. cashmiriana* dislikes lime and is particularly at home in this situation, especially in the north of England and Scotland where it does best. Like all corydalises it has swollen roots, and it dies down to below soil level in winter. Mark the site of the plants well so that cultivations do not disturb the tuber. Try growing the plant in a pot containing a mixture of one part John Innes ericaceous compost, one part peat and two parts sharp grit in the alpine house if you must, but remember not to let it become sun baked. Never let the compost become bone dry in winter, and repot only when absolutely necessary in spring. *C. transsilvanica* is most frequently grown as an alpine house plant – treat it as recommended for *C. cashmiriana*. *C. solida* is happy in a peat bed or a patch of peaty soil on the rock garden, and *C. cheilanthifolia* is best spurting from a crevice or the wall of a rock bed. It is unfussy as to soil.

Propagation Divide mature plants very carefully in spring, taking care not to damage the tuberous roots, nor to let them dry out.

Cotula (Compositae)

As a mat-former for use among paving or in alpine lawns, *C. squalida* (New Zealand) is a good choice. Its fresh-green leaves are small, feathery and finely cut and often tinged with bronze. The plant will never win prizes for spectacle, but it makes a good thick rug of foliage to show off its brighter neighbours.

Cultivation An easy plant for a sunny spot in any well-drained soil. Only sticky soil is not to its liking – a combination of moisture and shade will cause it to sulk, rot and die off.

Propagation Divide mature clumps in spring. Seeds can be sown in pots of seed compost in spring and germinated in a frame.

Crepis (Compositae) Hawk's Beard

These aristocratic dandelions are worth growing if you've space to spare. *C. incana* (Greece) has rosettes of downy, greyish leaves over which appear 9-in (23-cm) stems of pink dandelion flowers for most of the summer. *C. aurea* (European Alps) has orange flowers carried on 3 or 4-in (8 or 10-cm) stems, also in summer. Don't be snobbish about them – they'll both earn their keep.

Cultivation Bright sunshine and any well-drained soil will suit either plant, and *C. aurea* is a good choice for an alpine lawn.

Propagation Divide mature clumps in spring or autumn. Sow seeds in spring in pots of compost, and germinate in a garden frame.

Cryptogramma (Cryptogrammataceae) Parsley Fern

Although it's not included in most alpine books, I've sneaked in the parsley fern, *C. crispa* (Europe, including Britain, to the Caucasus and Siberia) because it looks so good when pushing up through a peat bed or the stony surface of a scree. It's deciduous and dies down in winter, so its fronds are especially fresh and feathery when they emerge in spring. It seldom grows more than 6in (15cm) high, and although it creeps it is unlikely to become a nuisance.

Cultivation This is a plant for acid soils, for it hates lime. Grow it in sun or shade in a peat bed or in a scree. Alternatively, grow it in pans in the alpine house, potting it up in a mixture of equal parts stone chippings (sandstone or granite) and sieved leafmould. Keep it cool at all times, and never let the compost dry out completely in winter.

Propagation Divide established plants in spring. Sow spores as soon as ripe as for *Adiantum*.

Cyclamen (Primulaceae) Sowbread

The charming hardy or near-hardy species of cyclamen win their way into any rock gardener's heart by virtue of their willingness to oblige and their handsomeness in both leaf and flower. There are many different species requiring slightly different techniques of cultivation, so instead of going into lengthy details for each of my selected species I have grouped them according to their requirements, and kept my descriptions down to colour, time of flowering and country of origin. Any cyclamen grower who wants to know everything about his plants should invest in a copy of *Cyclamen*, by D. E. Saunders, published by the Alpine Garden Society. It is cheap and unrivalled. Species in cultivation include:

C. africanum (Algeria), pale to dark pink, autumn, F.
C. balearicum (Balearics and southern France), white tinged pink, spring, F.
C. cilicium (Turkey), white to deep rose pink, autumn, F.
C. coum (From Bulgaria eastwards to Turkey, Asia Minor and Iran), white to deep carmine-red, winter through to spring.
C. cyprium (Cyprus), pale pink becoming white, autumn, F.
C. graecum (Mediterranean), pale to deep pink, autumn.
C. hederifolium (syn. *C. neapolitanum*) (Mediterranean), rich pink to deep pink, or white, summer to autumn, F.
C. libanoticum (Lebanon), white through pale pink to rich pink, late winter.
C. mirabile (Turkey), pale pink, autumn.
C. persicum (Mediterranean), white, lilac, pink, deep pink, early spring. F.
C. pseud-ibericum (Turkey, Asia Minor), magenta or lilac-pink, early spring, F.
C. purpurascens (syn. *C. europaeum*) (Europe from Austria to Bulgaria), pink to magenta-purple, summer to autumn and even early winter, F.
C. repandum (Mediterranean), carmine-red to pink and occasionally white, spring, F.

The species marked F are usually fragrant.

Cultivation *C. coum*, *C. hederifolium* and *C. purpurascens* can usually be grown outdoors with no protection from the weather. They enjoy best a spot in dappled shade rather than full sun, and prefer a good, loamy soil that has been enriched with plenty of peat and leafmould. *C. hederifolium* is quite at home in a peat bed. The remainder are best grown in deep pans plunged in sand in a frame, and brought into the alpine house (if necessary) when they are in bloom. The frame must be in a sheltered spot and capable of being shaded from bright sunshine. Grow all the species in a compost consisting of equal parts John Innes No. 1 potting compost, sieved leafmould and sharp grit. Both pot-grown and garden-grown plants must be set at the right level in the soil – the tuber should rest just below the surface, but the cluster of leaves and shoots should be just above. Don't adopt the old gardeners' method of planting so that the whole tuber is on top of the soil – some of these hardy cyclamen root from the top of the tuber around the leaves and will become starved and dehydrated if not bedded into the compost. Conversely, do not bury the tubers too deep or rotting of the leaves may result. Buy new tubers while they are in growth – tubers bought dry may be dead or at least too dry to survive. While they are growing all the species need plenty of water which can be applied to the surface of the compost as soon as it starts to feel dry. In very sunny weather shade the frame and ventilate well at all times except when frost threatens. In winter lag the frame at night (and during the day if frosts persist) with sacking or old carpeting, or even straw bales. Heavy frosts may kill the tubers. Air-warming cables controlled thermostatically to switch on when frosts occur will save the very tender species such as *C. africanum*, *C. cyprium* and *C. persicum*. During summer the following species should be allowed to dry out completely in their pots: *C. africanum*, *C. cyprium*, *C. graecum*, *C. libanoticum*, *C. persicum*. Put them in a sunny but well-ventilated frame or greenhouse as soon as their foliage has died down, and let them become sun baked. Start watering (gently at first) when growth recommences in autumn. *C. balearicum*, *C. cilicium*, *C. coum*, *C. pseud-ibericum* and *C. repandum* can be allowed to dry out considerably from late May to early July, but they do not need the baking preferred by the other species, and they should be prevented from drying out totally. Repot the tubers only when they are absolutely pot bound. The operation should be carried out immediately after the resting period, just before growth starts up. Tubers that do not need repotting can be top-dressed with the potting mixture at the same time of year. Scrape away about 1in (25mm) of the old compost and replace it with new, but take care not to damage the roots. Greenfly love cyclamen, but can easily be controlled with an insecticide containing pirimicarb. Vine weevils are fat, white grubs that eat away the roots and the tuber causing the leaves to collapse and die. These pests are lethal once they have established themselves in the compost. Watering the compost with a solution of HCH diluted to spray strength will control minor outbreaks; severe infestations mean death to a cyclamen.

Propagation Seeds should be sown as fresh as possible – straight from the split capsule if you can. Stored seed can be soaked in water for two or three days to soften the seedcoat. Sow in seed compost, lightly covering the seeds, and germinate them in a cold frame shaded from bright sunshine. Germination is often slow, and old seed may take up to a year to emerge. Do not allow the compost around the young plants to dry out at any time or they may become dehydrated. Leave them in the pot until they are so large that they restrict each other's growth, then pot them up individually.

Cymbalaria (Scrophulariaceae) Toadflax

Our now-native ivy-leaved toadflax, or Kenilworth ivy, *C. muralis*, is certainly a charmer when cascading down dry walls, but it's just too rampant to be let loose on most rock gardens. Grow instead its form 'Nana Alba' which is smaller, less rampant, and possesses small, white snapdragon flowers tinged with yellow. *C. hepaticifolia* (Corsica) is also modest in its territorial spread. Its kidney-shaped, greyish-green leaves show off well the lilac-purple flowers which, like those of the other species, appear in summer. *C. aequitriloba* (Corsica, Sardinia) is the most compact of the lot, making an inch-high (25-mm) rug of dark, evergreen leaves which are studded with minute, purple snapdragons in summer. (All the plants in this genus were once included in the genus *Linaria*.)

Cultivation These are excellent plants for establishing in dry walls and in the gaps between paving stones on a patio. They love to force their roots down into crevices, and will spread or trail their stems over most obstacles. *C. aequetriloba* is good in alpine lawns. Any ordinary soil that is well drained suits them, and a spot in full sun or dappled shade will be enjoyed. It makes sense to avoid planting cymbalarias of any kind next to rather feeble alpine treasures – you may never see your weaklings again.

Propagation Divide plants in spring or autumn. Sow seeds in pots placed in a garden frame in spring.

Cypripedium (Orchidaceae) Lady's Slipper Orchid

The lady's slipper orchid is to some gardeners what the Penny Black is to a stamp collector. The trouble is that the orchid is not so easy to keep, for it has definite likes and dislikes when it comes to cultivation. There are three species that are fairly freely available from nurserymen and bulb merchants. *C. calceolus* (Europe, including Britain) is most people's favourite. It grows to around 1ft (30cm) high, has broad, deeply ridged, fresh green leaves and chocolate-brown flowers with rich-yellow pouches. It blooms in early summer. *C. reginae* (North America) is less delicate in form (having a very swollen pouch) but more delicate in colour, being white flushed with pink. It flowers at the same time as the British native, but is rather taller. *C. acaule* (North America) is the ugliest of the bunch and has a pouch that looks like the nose of someone too keen on the bottle – inflated and veined with rose purple. The petals and sepals are greenish brown, and the plant seldom reaches more than 1ft (30cm) high. Although *C. calceolus* is a British native, there are now few places where it remains. It goes without saying that under no circumstances should plants be gathered from the wild.

Cultivation *C. calceolus* normally grows in woodland over limestone, but the addition of lime is not essential to its wellbeing – indeed it thrives in a peat bed if the aspect suits it. Give it a position in gentle shade which is sheltered from winds and bad weather. If you don't plant it in a peat bed, then enrich your soil with plenty of leafmould and peat to make a spongy, moisture-retentive medium. Plant the rhizomes so that they are just below the surface of the soil. Should the shoots emerge too early in spring, protect them from frost with a cloche. Alternatively, plant them in a north-facing bed so that they will emerge later. The other two species dislike lime, but rather than being planted in a peat bed they should be established in peaty, leafy soil, again in a sheltered spot (though not one riddled with tree roots). All three species die down in winter. *C. acaule* will tolerate a slightly drier soil than the other two. The plants can be grown in pans in a shaded frame which is well ventilated at all times except when frost threatens. Pot them up in a mixture of equal parts John Innes ericaceous compost, sieved leafmould and sharp grit. Only just cover the rhizomes with compost, and then top-dress with a ½-in (13-mm) layer of sieved leafmould. Water freely as necessary as soon as growth starts in spring, but slacken off when the foliage starts to fade after flowering in summer. When the foliage dies down keep the compost very much on the dry side, applying only sufficient water to prevent it from becoming dusty. Alternatively, the pans can be plunged in sand in a frame, and they will take up all the moisture they need if the sand is kept damp. Protect the plants from spring frosts and from early morning sun, which might scorch the new growths. Top-dress with a little fresh compost each spring – removing and replacing the leafmould before you do so. Repot only when absolutely necessary after flowering.

Propagation Divide mature clumps or potfuls (if you can pluck up the courage) in spring just before growth commences. Sections of rhizome with a shoot and some roots should be potted up or planted out individually. Keep the youngsters well shaded until they are established.

Cytisus (Leguminosae) Broom

Although it should strictly be relegated to the shrubs section at the back of the book, *Cytisus* is such an established plant on the British rock garden that it deserves more detailed treatment. The hefty hybrid brooms cannot be accommodated on the rock garden, but a handful of dwarfs can. *C. ardoinii* (Maritime Alps) demands your attention on several counts. First of all it is small – barely 6in (15cm) high; it spreads very slowly to no more than 1¾ft (45cm) across; it is long-lived and it smothers itself with bright yellow flowers in late May. *C. hirsutus demissus* (Greece) is another 'must'. The plant makes a flat carpet of growth – but not a rampant one – and its leaves are clad in silky, silvery hairs. The golden-yellow flowers are rather large for the size of the plant, and their keels are a rich russet brown. For the largest rock gardens *C.* X *beanii* and *C.* X *kewensis* are good choices, but they do need room to cascade their stems like floral waterfalls. The first is rich yellow and the second creamy white. Just one more for a change of colour: *C.* 'Peter Pan' is an upright grower to a couple of feet (60cm), and has rich-red flowers. Site it where you want a bit of height.

Cultivation Brooms relish hot, dry, sunny spots, and don't mind chalky soil (though they can be short-lived over very shallow, chalky earth). Plant them in autumn or late winter from pots to avoid root disturbance. These dwarf types need little pruning, but any straggly growths can be cut out immediately after flowering to encourage the production of young shoots. Do not cut back into really old wood, which will refuse to sprout.

Propagation Heel cuttings of young shoots just 3in (8cm) long can be rooted in sandy compost in a propagator in early summer. Firmer shoots can be rooted later in summer and autumn. Seeds can be sown in pots of compost placed in a frame in spring.

Daphne (Thymelaeaceae)

Plants with fragrance as well as colour are doubly desirable, and the daphnes are second to none when it comes to strength of perfume. There are a dozen or more species that are suitable for the rock garden but the following five are among the best. *D. cneorum* 'Eximia' is perhaps the most widely available. It is evergreen, and originally hailed from central and southern Europe, but the form 'Eximia' was selected by the late A. T. Johnson in his garden in Wales (if you've not read his *A Woodland Garden*, *The Mill Garden* or *A Garden in Wales* you're missing a treat – they're out of print

A dry and rocky bank at Hidcote Manor in Gloucestershire billowing with bright blooms of *Helianthemum, Cistus, Dianthus* and other sun worshippers in early summer.

Few primulas have more impact than the candelabra types. These are the 'Harlow Car Hybrids' providing an unrivalled spectacle in June at the Northern Horticultural Society's Gardens near Harrogate in North Yorkshire.

On rooftops and walls, succulent houseleeks or sempervivums are a bright foil for fussier alpines. Here they are bursting out of a stone trough and brimming with summer blooms.

You'll not find a dainter poppy than *Papaver alpinum,* with its finely cut grey foliage and nodding bells of pink, white, yellow or orange. It might be short lived but it sets plenty of seed.

A small collection of young, pot-grown alpines brighten my conservatory in spring. When the flowers fade they are moved to a garden frame and summer-flowering pot plants replace them.

A potful of *Rhodohypoxis baurii* is guaranteed to draw gasps of admiration when it bursts into bloom in spring or summer, but the plant can be equally spectacular out of doors.

Lumps of tufa grouped together make a home for a mixed collection of alpines at the Royal Horticultural Society's garden at Wisley.

Few plants offer such a striking display as *Tropaeolum polyphyllum,* with its finely cut blue-grey leaves and rich yellow June flowers.

Raised beds give alpines the sharp drainage they app
ciate. Here *Sedum kamtschaticum* 'Variegatum' and
other sun lovers grow happily in a concrete block be

These natural screes of the Wetterstein range in Bava
show the conditions in which many alpines thrive.
Their rooting mixture is almost pure stone chipping

but available from second-hand booksellers). On my chalky soil this beauty does very well, making a sprawling bush that is smothered with rich-pink, scented flower clusters in May. It seldom reaches more than 6 or 9in (15 or 23cm) high, but can spread to well over 3ft (90cm) – Johnson's plant was 6ft (1m 85cm) wide when it died. If you grow no other daphne, grow this one. Alpine freaks have enthused over *D. petraea* for years. It occurs in the southern European Alps and makes a flat mat of dark-green growth which shows off well the pink flower clusters when they appear in spring. In cultivation the form 'Grandiflora' is usually grown, though it is available only from a handful of nurserymen, who probably have waiting lists for it. Seldom offered on its own roots, it is usually grafted on to *D. mezereum*. Of the rest, *D. collina* (a plant of dubious origin and a name of dubious validity), *D. balgayana* (south-east Europe), and *D. arbuscula* (Carpathians) are all worth growing, and are offered from time to time by nurserymen. If you want to acquaint yourself more firmly with these and other members of the genus you'll enjoy *Daphne, the Genus in the Wild & in Cultivation* by Chris Brickell and Brian Mathew, published by the Alpine Garden Society. It's a book that combines scholarship with readability.

Cultivation 'It just died' is a phrase you'll hear about daphne more than any other. Sudden death is a fact of life with this genus, but often it can be traced back to some fault in cultivation, or to the plant flowering so well that it exhausted itself. As a general rule daphnes will do well on any soil, whether or not it contains chalk or lime. They need their shoots and leaves in bright sunshine, and their roots in a well-drained soil that is not likely to dry out in the heat of summer. Try them on a scree bed that offers a deep, cool, root run, or in a crevice between two rocks on the rock garden. My own soil is chalky, rather on the stiff side but very well drained, and it seems to suit daphne well, for it seldom becomes dusty except on the surface. Drought is a common cause of failure. On light and rather heavy soils work in plenty of peat or leafmould to aid drainage and make for some beneficial moisture retention. Top-dress the area around the plant with a little John Innes compost or even very well-rotted manure each spring, and make sure that plenty of water is given in periods of drought. When buying new plants, always choose youngsters with plenty of healthy, green shoots. Plant them in early spring, having made sure that the root ball was moist right through at the time of planting. Once established the plants resent disturbance – their thick roots travel extensively. Pruning is not, as a rule, necessary, but unwanted, wayward or unsightly branches can be cut out in spring. Large cuts should be treated with a fungicidal wound paint. The plants can be grown in pans in the alpine house to add fragrance and extra colour to the display. A mixture of equal parts John Innes No. 2 potting compost, sieved leafmould and sharp grit suits them, and they should be watered freely when necessary during spring and summer; rather more carefully in autumn and winter so that the compost is kept just moist. Repot only when absolutely necessary in spring.

Propagation The species mentioned can be propagated either by layering suitable shoots in March and April, and detaching these one year late when they are rooted – potting them up into the recommended mix – or by taking cuttings. Heel cuttings 2 or 3in (5 or 8cm) long will usually root in a heated propagator at any time between June and August. Experience will dictate the best time for each species.

Dianthus (Caryophyllaceae) Pink

Who could resist growing a few alpine pinks? Their flowers are dainty, bright and exquisitely scented, and they are just the plants for gardeners on chalk to grow in quantity. Dozens of different species, and twice as many hybrids, are offered by nurserymen, but one you must possess is *D. alpinus*. It's a European alpine that makes a spreading, flat cushion of narrow leaves that are topped with large pink flowers on 2-in (5-cm) stems in early summer. The blooms may be anything from pale pink to deepest rose, and usually have darker blotches around the central eye. Sadly there's no scent to this species, but then you can't have everything. *D. gratianopolitanus* (syn. *D. caesius*) is our native Cheddar pink, most desirable in its form 'Compactus'. The blooms are a lovely soft pink, and have that fragrance you're looking for. The plant grows about 4in (10cm) high. There are plenty of hybrids of *D. deltoides* (Europe), the maiden pink, but I find them singularly unimpressive. The blooms are brilliant, right enough, and carried in copious quantities, but at ½in (13mm) across they are rather too small for the length of stem – 6in (15cm) or so – and not in the slightest bit fragrant. They'll bloom between June and August. Far better are hybrids such as 'Pike's Pink', which makes tufted mats of blue-grey leaves, topped with little double pink flowers on 4-in (10-cm) stems, and is deliciously scented. And finally, a curiosity: *D. erinaceus*, the hedgehog dianthus. This cushion-former comes from Asia Minor, and makes tight domes of rather spiky leaves on which sit single pink flowers in early summer. Well suited, this plant will make a mound fully 2ft (60cm) across in time, and the stems are so close packed that the plant looks like a gigantic clump of moss.

Cultivation All pinks adore sun, and will sulk if they don't get it. Give them a patch of well-drained soil that is nevertheless unlikely to dry out completely in summer, and make sure that the ground is not too acid. Work in some limestone chippings if your soil is not naturally chalky. In sinks and troughs, retaining walls and even the crevices between paving stones, alpine pinks will thrive and enrich your summer evenings with their perfume. Carnation fly can sometimes attack.

Propagation Seeds can be sown in pots of seed compost in spring and germinated in a garden frame. Cuttings of young shoots can be rooted in a frame in summer. Mature, clump-forming plants can be divided in autumn.

Diascia (Scrophulariaceae)

This is a favourite of mine and an alpine that should be more widely grown. *D. cordata* (South Africa) makes a loose, spreading carpet of stems from which rise spires of warm-pink flowers, each with a large lower petal and two prominent spurs at the rear. The hybrid *D.* X 'Ruby Field' is even more remarkable, more floriferous and a little bit taller at around 9in (23cm). Its blooms are the colour of smoked salmon, and make a bright, long-lasting show from June onwards. The larger, but equally spectacular, *D. rigescens* has recently come on to the market and should soon win an army of fans.

Cultivation Grow *Diascia* where it has room to spread sideways on gently sloping ground, and then you'll see the display of blooms to best effect. Any ordinary soil that's not poor, dry and dusty will suit it, provided that drainage is good. It seems to tolerate chalky ground and can be clipped over after flowering. Slugs love it.

Propagation Divide clumps in autumn or spring. Cuttings can be rooted in a propagator in summer.

Dicentra (Fumariaceae) Dutchman's Breeches

Although most of the dicentras are plants for the flower border or woodland, there are one or two that happily add their grace to the rock garden. *D.* 'Stuart Boothman' commemorates a great alpine nurseryman and is a great plant. From its thick, carroty roots emerge feathery leaves of almost carroty fineness, but they are a rich blue-grey with Victoria-plum-coloured stalks. The late spring and early summer flowers are soft pink, and carried on 10-in (25-cm) stems, so that they dangle gracefully over the filigree foliage. It's a real beauty. So too is a relative newcomer that can be had from one or two sources. It's called *D.* 'Langtrees', and although it's only been with me for a season, I feel confident in recommending it. The foliage is steel-grey (as are the stalks), still finely cut, though not so filigree as 'Stuart Boothman', and the blooms are creamy white, flushed pink. At 8 or 9in (20 or 23cm) it's just a little smaller than its relation.

Cultivation They're a versatile bunch, the dicentras: they're at home in dappled woodland in a deep, leafy soil, or on the rock garden in full sun or dappled shade, so long as their roots can sink into well-drained but moisture-retentive soil that will not dry out in summer. Enrich the patch they are to occupy with plenty of leafmould or some peat and they'll soon settle in, even if your soil is chalky. They die down in winter, so avoid disturbing them inadvertently during their resting period. The buds that emerge in spring are soft, fleshy and brittle, so keep clear until they expand.

Propagation Divide mature clumps in spring as soon as the foliage has expanded a little. Avoid breaking the long, fleshy roots.

Dionysia (Primulaceae)

Every gardener has his Achilles heel – *Dionysia* is mine. At show after show in spring I admire these plants with their tight cushions of growth studded with yellow or pink flowers, and I vow to do better next year. As it is, I can just about manage *D. aretioides* (Iran) in its forms 'Gravetye' or 'Paul Furse', but I refuse to waste more money on the other difficult species until my competence with them improves. They are a *very* difficult group of plants to grow, unless you know the exact secrets of success, and have plenty of time to devote to them. That said, you might like to have a go with *D. aretioides* when you feel you've got the time and patience to spare. It's a beauty that makes a rounded dome of hairy rosettes which, with any luck, will be studded with stemless, bright-yellow primula flowers in March or April.

Cultivation *Dionysia* is for the alpine house or frame. I've seen it tried outdoors in tufa, but I've never yet seen a good plant that wasn't protected by glass. I grow my *D. aretioides* in a mixture of one part John Innes No. 1 potting compost and two parts sharp grit. At no time should the plant be overpotted. Water it in spring and summer whenever necessary (the cushion will feel slightly limp when water is required and the pot will feel light). I always check my pots for water by weighing them in my hand. After a while it's easy to tell whether the compost is moist or dry. *D. aretioides* can be watered from above, provided that care is taken in winter not to wet the foliage. In winter the compost should remain *almost* bone dry; the slightest trickle of water being given only when the compost is dusty and the foliage is becoming limp. Overwatering is the commonest cause of failure. Shade the plants lightly in summer and ventilate well in all but frosty weather. In winter the plants should be kept in an alpine house or frame, but they need no artificial heat – mine came through the winter of 1981/2 unscathed in an old cold frame. The foliage naturally turns brown in winter, and just the stem tips stay green. Watch out for greenfly on the shoots, especially in spring and summer. Repot only when absolutely necessary in March, and be prepared for disaster! Many growers keep the plant in quite a small pot, plunging this in a larger pot of sand. The sand is kept moist and the *Dionysia* extracts the water it needs. Keep the plants well shaded for a few weeks after potting. At any time, stems may die off; remove them cleanly as soon as they are noticed to prevent the rot spreading back. Do not remove the leaves that naturally turn brown in winter. The best of luck!

Propagation This is only for gamblers and experts. Seeds can be sown (when available) in pots on the surface of a mixture of John Innes seed compost and sharp sand. The seeds (or chaffy, leafy mixture in which the seeds are contained) should be lightly covered with grit. Sown in autumn, they are put in an unheated frame and allowed to be frozen during the winter. If you're lucky you might get some seedlings in spring; if you're not it may be anything up to three years before they emerge. Potting up the young plants is another dicey business. Some gardeners seem to succeed with cuttings consisting of single rosettes which are rooted in a closed frame in July. To discover more about the genus, read Christopher Grey-Wilson's *Dionysias: the Genus in the Wild and in Cultivation*, published by the Alpine Garden Society.

Dodecatheon (Primulaceae) American Cowslip, Shooting Stars

These elegant plants have upright rosettes of leaves and tall, straight flower stems that carry cyclamen-like blooms with reflexed petals. There are a number of species that vary little in their appearance, but the two most commonly available are *D. meadia* (north-east America), 1ft (30cm), deep rose pink; and *D. pauciflorum* (syn. *D. pulchellum*) (north-west America), 9in (23cm), lilac pink. Both flower in early summer.

Cultivation These are good plants to establish at the side of a stream or pool where they can sink their roots into a rich, moisture-retentive earth. They enjoy best dappled shade, but can tolerate full sun provided the soil holds plenty of moisture. Stiff pockets of earth on the rock garden may keep them happy – work in plenty of leafmould.

Propagation Divide mature clumps in spring or autumn. Sow seeds in pots of sandy compost placed in a cold frame in spring.

Doronicum (Compositae) Leopard's Bane

In spite of the fact that I've seen them growing in the Alps, doronicums seem to me to be better off in the herbaceous border than on the rock garden. *D. cordatum* (Europe and western Asia) is the most suitable for the large rock garden, where it will produce plenty of bright-yellow daisies in early spring over a mound of rounded, fresh green leaves. It grows about 6in (15cm) high.

Cultivation A sunny spot in any well-drained soil suits them.

Propagation Divide clumps in autumn.

Douglasia (Primulaceae)

These neat little cushion-formers are closely related to *Androsace*, and frequently confused with them by both gardeners and botanists. *D. vitaliana* (Spain and European Alps) is the species most widely grown and most easily obtained, especially in its form *D. v. praetutiana*. It makes a small, fine-needled cushion of greyish green, which is studded with acid-yellow, stemless flowers in early summer. *D. laevigata* (North America) has deep, rose-pink blooms on its cushion in May or June, and *D. montana* (North America) is similar but of rather beefier stature.

Cultivation Grow in bright, sunny spots on a scree bed, in sink and trough gardens or crevices between rocks; always where drainage is excellent. They don't mind chalky soil. A pane of glass will offer welcome protection in very wet winters. Alternatively, grow them in an alpine house or frame in pots containing a mixture of equal parts John Innes No. 1 potting compost and sharp grit. Plant or repot in spring when necessary. Container-grown plants should be watered well when dry in spring and summer, rather more tentatively in winter, but the compost should never be allowed to dry out completely. Shade from brilliant summer sun when grown in pots.

Propagation Sow seeds in pots of sandy seed compost in spring, and germinate in a cold frame. Divide mature cushions carefully in August, and pot up rooted portions for growing on. Keep in a shaded frame until established.

Draba (Cruciferae) Whitlow Grass

There are dozens of drabas – some rather dull and others bright treasures worth growing wherever you've got the space. All the following kinds are among the treasures: *D. aizoides* (Europe, including Britain), tufted rosettes carrying yellow flowers on 2-in (5-cm) stems in spring; *D. bryoides imbricata* (Caucasus), densely packed cushions of minute rosettes, clusters of yellow flowers on thin, fusewire stems in spring; *D. mollissima* (Caucasus), downy rosettes making firm hummocks, yellow flower clusters are carried on 2-in (5-cm) stems in spring; *D. polytricha* (Armenia and eastern Europe), plump hummocks of growth topped with yellow flowers on 2-in (5-cm) stems in spring. The last two are the best but the trickiest.

Cultivation *D. aizoides* and *D. bryoides imbricata* can be risked outdoors in troughs and sinks or sunny spots in rock gardens or beds. *D. mollissima* will often do well outdoors in tufa (if you can prevent the moss from taking over), but it, and *D. polytricha*, are most reliably grown in the alpine house or frame. Pot them up in a mixture of equal parts John Innes No. 1 potting compost and sharp grit. Water them freely when they are dry in spring and summer (keeping water off the cushions), but more sparingly in winter so that the compost is very much on the dry side. Remove rotting rosettes as soon as they are seen (small pieces of stone can be wedged under gappy rosettes to tighten them up). Repot annually after flowering.

Propagation Seed can be sown in February or March in pots of sandy seed compost placed in a cold frame. Individual rosettes can be induced to root if inserted in a propagating frame in early summer. Mature plants can be divided, and rooted portions potted up individually in spring or early summer.

Dryas (Rosaceae) Mountain Avens

These are superb woody carpeters with leaves roughly oak shaped, dark, glossy green above and grey beneath. The flowers are lovely anemones of creamy white. *D. octopetala* (Europe, including Britain) is the commonest, and grows but 3 or 4in (8 or 10cm) high in a mat that will spread for a yard (metre) or more. *D.* × *suendermannii* is a hybrid between *D. octopetala* and the North

American *D. drummondii*. It inherits from this last-named parent the habit of carrying its flowers so that they nod bashfully in the breeze. All bloom in early summer and have feathery seedheads.

Cultivation Find for *Dryas* a bright, sunny spot where it will have room to romp. The soil needs to be well drained, but enriched with plenty of peat or leafmould. A scree bed will suit it well, and it is a good alpine lawn plant. It is blissfully happy over chalk. There is a smaller form of *D. octopetala* known as 'Minor', and this can be trusted not to overrun troughs and sink gardens (for a few years at any rate!). 'Minor' can also be grown in pans of gritty compost in the alpine house. Water it freely in spring and summer, more sparingly in winter and top-dress it each spring with the potting mix – equal parts John Innes No. 1 potting compost and sharp grit. Repot when necessary after flowering.

Propagation Sow seeds in pans of seed compost in spring, and germinate in a frame. Divide mature plants in spring. Take stem cuttings 3in (8cm) long with a heel of older wood in June, and root in a closed frame. Division is certainly the most reliable method – the other two may yield a low percentage of success.

Edraianthus (Campanulaceae)

Some gardeners think of these plants as *Edraianthus*, others as *Wahlenbergia*, but here I've grouped them all together for convenience. There are just two species you'll come across with any frequency, and both of them are worth having. *E. pumilio* (Dalmatia) wins by a nose in the beauty stakes on the grounds of its compactness as much as anything. It makes a tidy mat of pointed-leaved rosettes from which spring masses of tubular campanula flowers of lavender blue in May and June. *E. serpyllifolius* (syn. *Wahlenbergia serpyllifolia*) (Balkans) is rather looser in habit, though still ground hugging, and its bells are rather more open, with petals that turn under atttractively. The colour is similar, and the flowers usually open in June. There is a larger-flowered version called *E. s.* 'Major' which is offered by one or two nurserymen, though some stocks appear to be losing their vigour.

Cultivation Both plants love sun and a chance to send their roots deep into the soil – a limestone scree is most to their liking, but they'll also enjoy a well-drained patch on the rock garden or gritty soil in a sink or trough. The plants can be grown in the alpine house or frame in a compost consisting of equal parts John Innes No. 1 potting compost and sharp grit with some limestone chippings added. Water freely when necessary in spring, but much more carefully in winter, keeping the compost very much on the dry side. Repot every other year in spring. Soggy leaves in winter will quickly rot and the entire plant may die. Outdoors the protection of a sheet of glass will be beneficial in really wet weather.

Propagation Seeds can be sown in pots of compost as soon as they are ripe. Germinate them in a cold frame. Cuttings made from single rosettes can be rooted in a propagator in late April or May.

Epimedium (Berberidaceae) Bishop's Hat, Barrenwort

It is the leaves of these elegant yet robust plants that give them their common name – the bishop's mitre may be a little more symmetrical, but the overall shape is the same. There are at least half a dozen different species widely available, varying in their origins from Europe to North Africa, Iran and Japan, and their dangling, cross-shaped flowers may be white, pink, rose-red, yellow or violet. I group them all together because I think they are all meritorious, whether they reach 9in (23cm) or 18in (45cm). The flowers are very pretty but hardly long-lived enough to warrant their being the main attraction. This honour belongs to the leaves – fresh green or burnished deep-red on unfolding, turning through mid-green to russet brown as they fade in winter. If you feel you must go to your nursery armed with a few names, then I offer you *E. alpinum*, 9in (23cm), dullish red; *E. pinnatum*, 1ft (30cm), yellow; *E.* X *rubrum*, 1ft (30cm), red and white. All bloom in May.

Cultivation Epimediums are versatile plants. I've grown them on dusty sand (where they reluctantly manage to cling on to life, spreading very slowly) and on stiff soil well laced with organic matter (where they thrive and multiply). Any ordinary, well-drained yet moisture-retentive soil grows them, and they are happy too in the peat bed. Sun or shade suits them, and they don't care a fig about chalk, so long as some organic nourishment has been worked into the soil. Clip off the dead foliage quite close to the ground in February, or you'll miss the flowers when they appear in spring.

Propagation Divide clumps in autumn.

Erigeron (Compositae) Fleabane

These are dapper dwarf daisies – if you choose wisely. Some are weeds no better than dandelions. *E. borealis* (syn. *E. alpinus*) (northern Europe, including mid-Scotland), makes tufts of downy leaves from which rise up lavender daisies centred with yellow. The blooms are carried on 6-in (15-cm) stems in summer. *E. aureus* (North America) is a dainty miniature up to 3in (8cm) high, and its blooms are bright yellow. *E. mucronatus* (Mexico and now Europe) is the most rampant of my little bunch. It grows to around 9in (23cm), and throws up plenty of blooms varying in colour from white to rich pink from late spring to autumn.

Cultivation All daisies love sun and the erigerons are no exception. Site them in full light where the soil is very well drained and they'll thrive. Scree conditions are much appreciated, but you'd be well advised to keep *E. mucronatus* in a wall of the bed, or a separate dry wall, so that it does not overrun more respectable plants. It doesn't mind chalky soil. *E. aureus* is sometimes grown in pans in an alpine house or frame, in which case give it a compost consisting of equal parts John Innes No. 1 potting compost and sharp grit. Water freely when necessary in summer, but keep the compost very much on the dry side in winter. Remove any dead leaves and faded stems to prevent rot setting in.

Propagation Divide clumps in autumn or spring. Sow seeds in pots of seed compost placed in a garden frame in spring.

Erinacea (Leguminosae) Hedgehog Broom

This spiky thicket of a plant should be in every enthusiast's collection, for it is a remarkable sight. *E. anthyllis* (syn. *E. pungens*) comes from Spain, Algeria and the Pyrenees, and in cultivation will make a mound of spines 9in (23cm) high and perhaps 1ft (30cm) across, but only after a good many years. The stems (for seldom are there many leaves in evidence) are grey-green in appearance, thanks to a coating of silvery hairs. In May pale violet-blue pea flowers appear throughout the thicket, making a bright show.

Cultivation The plant is most reliably grown in the alpine house, where a pan of John Innes No. 1 potting compost mixed with an equal amount of sharp grit will suit it fine. Try it outside only on a sunny scree bed, for unless it has bright light and phenomenally good drainage it will rot off. Severe winters will kill it, even in an alpine house or frame, but the gamble is well worth taking. Water pot-grown plants freely in spring and summer, more carefully in winter when soggy compost means death. A good baking in late summer (when the compost can be left dry for longer periods between watering) will ripen up the stems to initiate next year's flowers.

Propagation Sow seeds, if you can get them, in pots of seed compost in February, and leave in a garden frame for a month before bringing into a warm greenhouse to germinate. Prick out the seedlings when young to avoid root disturbance. Some gardeners recommend taking cuttings of young shoots with a heel of older wood immediately after flowering. These apparently root if inserted around the edge of a pot of very sandy compost which is plunged in a garden frame. The young plants will be ready for potting up (if all goes well) by September.

Erinus (Scrophulariaceae) Fairy Foxglove

Figure 2.1
Erinus alpinus
Fairy Foxglove

Alpine gardeners who like their plants neat will thoroughly approve of *E. alpinus* (western Europe) (fig. 2.1) which makes smart tufts of slightly downy leaves with indented edges. Each hummock erupts into bloom in May and June, when the rose-pink flowers open in a dense spike atop each stem. There is a white form, and the blooms may also be deeper coloured on some plants. All grow to around 4in (10cm).

Cultivation Plant in sunny spots on the rock garden, in troughs and sink gardens, in crevices between rock or in walls and paths. Any well-drained soil is enjoyed, even if chalk is present. Spring is the best time to plant.

Propagation Divide clumps in early spring. Plants can be short lived, but are easily renewed from seeds. These can be sown in pots of seed compost in a frame in spring, or direct sown, also in spring, where the plants are to grow.

Eriogonum (Polygonaceae)

Not many of these choice plants are freely available, but *E. umbellatum* (western North America) is an exception well worth growing. It makes deep mats of leathery leaves that are grey-green in summer and red tinged in winter when the frosts burnish them. From the middle of summer onwards, pale-yellow flower clusters appear on 6-in (15-cm) stems to offer lasting enjoyment.

Cultivation Although most eriogonums insist on alpine-house treatment, this one doesn't. It will be happy in well-drained soil on the rock garden or in a scree bed or a trough of gritty compost. Unlike many of its relations it will put up with chalky soil, and it thrives in sun. Winter wet is not much to its liking, so a pane of glass propped over it will nurse it through the foul weather. If other species come your way grow them in pans in the alpine house or frame. A potting mix of equal parts John Innes No. 1 potting compost, sieved leafmould and sharp grit will be enjoyed. The plants like a free supply of water during the growing season, but need to be kept much drier (though not desiccated) in winter. Protect them from rain and from overhead drips. Repot only when the existing container is outgrown – spring is the best time for this.

Propagation Take cuttings of young shoots with a heel of older wood during summer, and root in a garden frame in a gritty medium. Seeds can be sown in pots of seed compost in a garden frame in spring, or as soon as they are ripe, if you can collect them from your own plants.

Erodium (Geraniaceae) Heron's Bill

The heron's bills are in many respects similar to their close relations the cranesbills (geraniums), but from the gardener's point of view they are usually rather more wiry with a touch more elegance. *E. chamaedrioides* (syn. *E. reichardii*) (Majorca) is the all-time favourite, and rightly so, for it carries plenty of white, pink-veined flowers on 2-in (5-cm) stems over its dome of foliage from early to late summer. There is a pinker variety 'Roseum' which is equally desirable, and a good white – 'Album'. *E. macradenum* (Pyrenees) is a touch bigger at around 4 to 6in (10 to 15cm), and has lilac-pink flowers blotched with two heavy spots of violet on the upper petals. The leaves are different too – carroty and grey-green. There are plenty of hybrids that are worth a try.

Cultivation Erodiums love a sun-baked spot, provided it does not turn to dust in the summer heat. Site them where their roots can run through any well-drained soil and they will romp away. In shade they will sulk and become spindly. They are happy in scree beds, and the smaller species will not take up too much room in spacious troughs or sink gardens.

Propagation Sow seeds as soon as they are ripe in pots of seed compost germinated in a garden frame. Cuttings consisting of single tufts of growth with a piece of older wood at the base can be rooted in a propagator in late spring and early summer. Rooted portions can be removed and transplanted from mature clumps in spring.

Erysimum (Cruciferae) Alpine Wallflower

Many of the *Crysimums* are annuals with brief lives, but there are a few that are reliably perennial, even if they do have to be regularly renewed from seed or cuttings to keep them in good shape. *E. linifolium* (Spain) is the most freely available species, and carries lilac flowers on stems 9 or 12in (23 or 30cm) long in summer. There is also a variegated variety, and I'm very partial to 'Bowles' Purple Wallflower', which is closely related, though rather beefier and perhaps more at home in the flower border. There are plenty of bright varieties to choose from: 'Sprite' grows to just 4in (10cm), and has primrose-yellow flowers; 'Moonlight' is similarly coloured, and 'Orange Flame' rather richer. All bloom from early to mid-summer. Some nurserymen lump them with *Cheiranthus*.

Cultivation Alpine wallflowers are easy to grow. Give them a sunny spot on the rock garden, or plant them in gaps left in paving. Any well-drained soil is tolerated, and chalky ground doesn't worry them at all.

Propagation Increase is easy – simply root shoot-tip cuttings about 2 or 3in (5 or 8cm) long in a propagator in spring or summer. Seeds of the true species can be sown in pots in a cold frame in spring or summer, or even in the open in shallow drills. Some plants will sow their own seeds and save you the bother.

Euphorbia (Euphorbiaceae) Spurge

There's just one euphorbia that I'd recommend for the rock garden, bank or wall and that's *E. myrsinites* from southern Europe. Planted where its stems have the chance to cascade down a vertical rock face, it will show itself off to perfection. The stems carry serried rosettes of blue-grey leaves, and are topped with acid-yellow flowers in summer – not the most subtle combination but quite eye-catching. The plant won't grow very high, but it does spread for a couple of feet (60cm) or more, so give it room to grow. Beth Chatto offers a clone labelled simply 'Good Form', which has larger flowerheads tinged with apricot.

Cultivation Bright sun is essential, and well-drained soil of any kind is suitable. Plant this spurge at the top of a wall or above a rock face or step so that it can travel downwards. It's good in the gaps between paving slabs too, but it's rather fleshy, so keep your feet away from it. If children roam the patio then this plant is not a good choice – its milky sap is highly toxic to both eyes and mouth. Cut away any dead or tatty stems in spring.

Propagation Cuttings of young shoot tips can be rooted in sandy soil in a frame in summer. Older plants can be divided in autumn.

Euryops (Compositae)

A smashing foliage plant that used to be thought of as too tender for outdoor culture, *Euryops* is now showing that it's hardy in all but the most severe winters. By far the best species, and the most popular too, is *E. acraeus* (syn. *E. evansii* of gardens). It comes from the Drakensberg Mountains of South Africa, and makes a dense, much-branched shrublet up to perhaps a foot (30cm) high. The leaves are long, thin, and slightly curved upwards, and each one is scored longitudinally. But it's the rich, metallic silver-grey colouring that gives the plant its charm, plus the bright-yellow daisies that it carries in summer. Slugs may nibble it.

Cultivation Well worth chancing outdoors anywhere in the country, give it a bright and sunny spot in sharply-drained soil, and if you're really nervous protect it in winter with a pane of glass. If you're idle, leave it alone. It will enjoy growing in a scree bed. Pinch out the shoot tips in the early stages to encourage a bushy habit. Some growers keep it in pots in an alpine house or frame, but I found that under such circumstances its lower leaves died off, giving it a very sorry appearance. If you must grow it in pots, use a mixture of equal parts John Innes No. 1 potting compost and sharp grit, and give it plenty of water when necessary in the growing season, much less in winter, keeping the compost almost dry. Annual repotting in spring will prevent starvation and leaf browning to some degree.

Propagation Cuttings of shoot tips can be rooted in a propagator in summer. 'Runners' are often produced around the plant, and these can be removed and planted in spring.

F

Festuca (Gramineae) Fescue Grass

These compact grasses can be grown on the rock garden with little fear of their becoming too rampant. Quite a few are on nursery lists, but perhaps the brightest is *F. ovina* 'Glauca' which makes big fat 'shaving brushes' of blue-grey, needle-like leaves. The species itself is widely distributed in Europe, but the variety is far more ornamental and grows to around 6 or 9in (15 or 23cm).

Cultivation The blue fescue grass is not difficult in any well-drained soil and a sunny spot. Where conditions are very moist it can rot off at the base, and in shade it will become spindly – falling open and losing its neat habit. It is quite at home in pockets of earth between paving stones, and is a natural tussock-former in the alpine lawn.

Propagation Divide mature clumps in spring.

Frankenia (Frankeniaceae) Sea Heath

Carpeters are indispensable on the rock garden, especially as edgers, and *F. thymifolia* (Spain) is a good choice. It makes flat mats of tiny, dark-green leaves, spreading sideways for 2ft (60cm) or more. Clusters of small, pink flowers stud the carpet of leaves in summer.

Cultivation *Frankenia* is a sun worshipper that will thrive in any light, well-drained soil. Work in some sharp grit or sand on heavier loams to improve drainage and friability. The plant is a natural choice for an alpine lawn, a sunny scree or gaps between paving stones on a patio. Keep it out of a sink garden, where it will swamp everything in sight. Spring-flowering bulbs can penetrate its canopy with ease and brighten up the foliage early in the season.

Propagation Divide plants in spring or autumn, and replant rooted portions.

G

Galax (Diapensiaceae) Wand Plant, Carpenter's Leaf

A woodlander handsome in leaf and flower, *G. aphylla* (syn. *G. urceolata*) is the only member of its genus. It comes from the wooded mountain slopes of North America, and makes mounds of glossy, rounded, leathery leaves that are usually rich green, but often burnished with rosy red. Foot-high (30-cm) spikes of tiny, white flowers appear in early summer, and although they are few in number they add greatly to the plant's attraction.

Cultivation This is a plant for the peat bed, where its roots can ramble happily through cool, moist, leafy earth. It resents chalky soil, and prefers some shade. Don't waste your time and money by planting it in a soil which is likely to dry out in summer. Colonies of the plant will make attractive, evergreen ground cover. It is worth trying in a moist, shady spot on the rock garden where the soil can be enriched with peat and leafmould.

Propagation Clumps can be divided and replanted at a spacing of 1ft (30cm) in spring or autumn. Seed, when available, can be sown in pots of peaty compost in spring, and germinated in a garden frame.

Genista (Leguminosae) Broom

Near relations of *Cytisus*, the genistas are a motley bunch of plants with large and small, carpeting and shrubby species. Choose carefully the ones you admit to your rock garden, for many of them will swamp more modest alpines. That said, the ones that will fit into the rock garden are treasures, especially in spring, when most of them wreathe their stems in bright pea flowers. *G. sagittalis* is my personal favourite. It's a European species that makes a thick rug of winged stems, tipped with bright-yellow flowers in summer. *G. lydia* (Balkans and Asia Minor) is larger, and should be used as a floral waterfall only on a sizeable rock bank where its 2-ft (60-cm)-long stems can arch and display the bright-yellow flowers to best effect. It's a late spring flowerer. *G. pilosa* 'Procumbens' is another spring bloomer, this time making more compact mats, and *G. hispanica* 'Compacta' is a prickly mound-former growing to around 1½ft (45cm). Both have yellow flowers.

Cultivation Full sun and a well-drained soil are the two essential requirements of all brooms. Genistas do not mind chalky soil, provided it's not too poor and shallow, but they resent root disturbance once planted, so choose the right spot for them from the start. If you've a dry bank that is inhospitable to most alpines, try genistas there. The mat-formers look best cascading over sheer rock faces or retaining walls, and they make bright additions to alpine lawns. Any pruning felt to be necessary should be carried out immediately after flowering, when some of the older, flowered stems can be removed to make way for the new. Avoid cutting hard back into really gnarled wood which might not sprout. The shrubby types can become scraggy after six or eight years, and are then best replaced with youngsters. Careful annual pruning will extend their lives.

Propagation Stem cuttings 3in (8cm) long can be made from shoot tips in late spring and early summer, and rooted in sandy, peaty soil in a garden frame or propagator. Rooting may be a little slow. Seeds can be sown in pots of seed compost, and germinated in a garden frame in spring.

Gentiana (Gentianaceae) Gentian

There's a certain magic about gentians, and nothing compares with that moment on a mountain climb when you come face to face with that first deep-blue flower shining from the dew-laden grass. There are hundreds of species in the wild, but just a handful of them are

Figure 2.2
Gentiana acaulis
Trumpet Gentian
(left)
Figure 2.3
Gentiana verna
Spring Gentian
(right)

offered by nurserymen, plus a generous helping of hybrids that have arisen in cultivation. There are basically two types of gentian – the European species which flower in spring and which will, in general, tolerate lime; and the Asiatic species, which are usually autumn flowering and need an acid soil. I've restricted myself to describing just six (not an easy task!) which are easy to grow given the right site and soil. *G. acaulis* (Europe) (fig. 2.2) is something of an enigma both in name and cultivation. The name apparently covers a number of distinct species, but to most gardeners the typical trumpet gentian has narrow, glossy leaves of light green and trumpets of glowing blue, spotted in the throat. It is barely 3in (8cm) high and blooms in spring. *G. asclepiadea* is the European willow gentian. It is a tall beauty with arching stems clad in willow-like leaves, and the flowers emerge from the leaf axils in summer. Give it space, for it grows to around 2ft (60cm). *G. septemfida* (Asia Minor), Caucasus, Iran and Turkestan) should be your first gentian because it's one of the easiest to please. It makes clumps of 6-in (15-cm)-high, green-leaved stems which erupt into blue trumpets in late summer and autumn. *G. lutea* (Europe) is the giant of the race and the plant from which tonic bitters are made. Only where you have 4ft (1m 25cm) of vertical space to spare should you grow this hefty plant with its oval, deep-ribbed leaves and yellow flower spike. *G. sino-ornata* (China and Tibet) is a classic example of the lime-hating Asiatic species. It makes a rosette from which long stems clad in narrow leaves are sent out to explore. The Brin form is one of the best, and will even scramble through rock garden shrubs to open its deep-blue trumpet flowers above the canopy in autumn. Finally, *G. verna angulosa* (fig. 2.3), which is a dainty, elfin gentian having a wide distribution through Europe and Asia Minor and forming a low plant with mid-green, oval leaves. Starry blue flowers open atop 3-in (8-cm) stems in spring.

Cultivation Each of the plants I've described has rather different requirements, so I'll tackle each one briefly on its own.

G. acaulis will grow in any well-drained soil with or without chalk. Sometimes it refuses to flower, in which case shift it around annually until it seems best suited, or try scattering a little tomato fertiliser around it, or a little lime if your soil is rather acidic. It likes best an open, sunny spot, and can be grown in scree beds or sinks, though some growers swear that fairly heavy soil and firm planting make for best results. At best it's a temperamental, if glorious, plant.
G. asclepiadea needs gentle shade and a deep, moist root run. It is very much at home in a peat bed.
G. lutea likes its foliage in the sun and its roots in a deep, rich, moisture-retentive soil. It takes a few years to establish itself and resents root disturbance.
G. septemfida is an accommodating plant. In any ordinary, well-drained soil, even a chalky one, it will do well, and it doesn't mind a tiny bit of shade, though full sun will produce the best plants.
G. sino-ornata insists on a lime-free soil, and repays a little effort in soil preparation. Work in grit, peat and leafmould to produce a moisture-retentive but well-drained medium. Shade from scorching noonday sun will be appreciated. Regular division every couple of years will keep the plant in good heart. The plant naturally dies back to its crown in winter.
G. verna angulosa is at its best when one or two year's old; after that it becomes rather straggly and not so attractive. Buy young plants, which establish themselves well. It will thrive in sinks, screes and troughs, and it loves lime. A sunny spot in gritty, leafy soil is much to its liking. Both it and *G. acaulis* can be grown in pots of equal parts John Innes No. 1 potting compost and sharp grit with a sprinkling of extra chalk or a few limestone chippings. Water them freely

in spring and summer when necessary, and keep them gently moist in winter – not soggy. Repot annually in early spring.

Propagation *G. asclepiadea*, *G. septemfida*, *G. acaulis* and *G. sino-ornata* can be propagated by division in early spring, as can *G. lutea* (though it resents disturbance). *G. verna angulosa* is usually propagated from seed sown as soon as ripe, or in spring in pots of seed compost placed in a frame; but short stem cuttings can be rooted in moist, sandy compost immediately after flowering. Don't let either the seedlings or the cuttings become scorched at any time.

Geranium (Geraniaceae) Cranesbill

The geraniums are among the best rent-payers you'll ever have in your rock garden – they flower for most of the summer and seldom get out of hand. Many geraniums are strong-growing border plants, it's true, but select the dwarfer species and varieties for rock beds, banks and patios, and you'll have no complaints about their vigour. *G. sanguineum lancastriense* is a native of the Lancashire coast and a fine garden plant. It makes carpets of finely cut, dark-green leaves topped with wide-faced, white flowers veined with blush pink. It's as easy as pie to grow, as is *G. cinereum* (Pyrenees). This species is clump-forming, and most widely grown in its variety 'Ballerina', and rightly so, for the parsley-cut foliage is topped all summer long with blooms of lilac pink, densely veined with rich magenta-purple. The variety 'Apple Blossom' should suit you if you are searching for a softer shade. *G. dalmaticum* (Yugoslavia and Albania) is the plant to choose if space is really in short supply, for it makes obsessively tidy clumps about 4in (10cm) high. The leaves are red tinged and the flowers pink or white, shaded from the centre with blush pink in the variety 'Album'. They don't stand the weather as well as the blooms of the previous two varieties, but plenty more are produced to replace them. *G. renardii* (Caucasus) is verging on the large side for the rock garden, but even though it reaches 9 or 10in (23 or 25cm) in height, it stays in a tidy clump and so sneaks in happily. The dome of grey-green, felted leaves shows off well the flowers, which are white, veined with deep purple.

Cultivation Sunny spots and ordinary, sharply drained soil produce the best plants, and they'll happily put up with chalky earth. Site the carpeters where they will have room to spread, and tuck the clump-formers into bright crevices, ledges or gaps between paving stones.

Propagation Divide mature clumps in autumn or spring.

Geum (Rosaceae) Avens

Most geums belong in the flower border, but one or two are compact enough – and from the right geographical areas – to be at home on the rock garden. *G. montanum* (European Alps) makes sizeable tufts of deeply cleft leaves from which spurt 4 to 6-in (10 to 15-cm) stems carrying wide-faced, golden flowers in early summer. I hesitate to mention *G. reptans*, which I've never grown and which is apparently a tricky customer. It comes from the European Alps, and is of a habit that would convince you it was easy to grow – it throws out runners and has plenty of yellow flowers which sit singly on 6-in (15-cm) stems. But it's miffy, and dies rapidly if not suited.

Cultivation *G. montanum* is at home in any decent soil that is well drained yet not dusty. It likes sun, but can put up with dappled light. *G. reptans* grows best in a lime-free scree, where its extensive root system can sink itself deep among a stony mixture. If you want to grow the plant on the rock garden, Clarence Elliott speaks of planting the roots in a 3-in (8-cm) layer of sand laid over light loam, and then piling a layer of stones around and over the plants so that just the tips of their leaves are visible – then they will thrive and push up to the surface while their roots remain cool and deep. You can try the same method in large pots in the alpine house or frame. Keep the compost slightly moist through the winter, and give plenty of water when necessary during the growing season.

Propagation Divide clumps or remove rooted runners in late spring. Seeds can be sown in pots of seed compost in spring, and germinated in a garden frame.

Globularia (Globulariaceae) Globe Daisy

Globe daisies always make me smile – probably because the fluffy, lavender-blue spheres of flower look as though someone has deliberately stuck them all over the mat of foliage to brighten it up. *G. cordifolia* (central and southern Europe) is the commonest species available. Its paddle-shaped leaves are tiny, dark green and notched at the tips, and its fluffy flowers sit on 2-in (5-cm) stems in summer. *G. repens* is similar but smaller at only 1in (25mm), and it hails from south-west Europe. You'll get something similar if you buy *G. bellidifolia* (now correctly known as *G. meridionalis*). *G. trichosantha* (Balkans) is larger – up to 8 or 9in (20 or 23cm) – and I think lacks the charm of the miniatures. It lacks their comical appearance too.

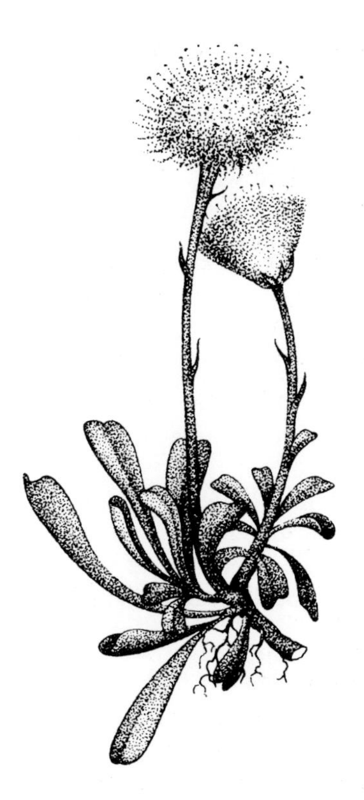

Figure 2.4 *Globularia cordifolia* Matted Globularia

Cultivation Grow the globe daisies in brilliant sun and in well-drained soil – they love chalk. Scree conditions will suit them well, as will pockets of soil in a patio, and they are fine components of an alpine lawn – the carpeting species, that is. The miniatures also make good pan plants in the alpine house or frame. Grow them in a mixture of equal parts John Innes No. 1 potting compost and sharp grit with a few limestone chippings added. Water them freely when necessary from spring to autumn, and keep them just slightly moist in winter. Repot after flowering every second year.

Propagation Divide clumps in spring or take short stem cuttings in summer, and root in a garden frame in sandy soil.

Gypsophila (Caryophyllaceae)

Unlike the border gypsophilas, the species mentioned here are carpeters. *G. repens* (European Alps) is the commonest species on offer in several forms. The one sold as 'Fratensis' is neat of habit and has rich pink flowers, and 'Dorothy Teacher' has markedly glaucous foliage which shows off the pink flowers a treat. 'Dubia' is another common form with pink flowers and leaves tinged bronze.

Both this variety and 'Fratensis' are really old synonyms for *G. repens*, but the plants sold under their descriptions do differ. White forms are also available. *G. cerastioides* (Himalaya) is a much neater creeper with tufts of felted leaves and summer flowers that are white, veined with purple. It will make just 2 or 3-in (5 or 8-cm)-high clumps, whereas *G. repens* can reach 6in (15cm), and is much more sprawling.

Cultivation These are plants for bright sunny spots among paving, in walls, cascading over rocks and, in the case of *G. cerastioides*, for scree beds. All gypsophilas can cope with and even enjoy chalky soil, and they insist on sharp drainage. Snip off the flower stalks once the blooms have faded.

Propagation Both species can be divided in spring or autumn, or seeds can be sown in pots in a garden frame in spring, or outdoors at the same time. *G. repens* can be propagated from early-summer cuttings rooted in sandy soil in a frame.

Haberlea (Gesneriaceae)

There are not many nurseries that offer *Haberlea*, but it's such a useful plant and one of considerable beauty that it must be mentioned. There are just two species: *H. rhodopensis* and *H. ferdinandi-coburgii*, both from the Balkans. They differ only slightly in that the second is rather larger flowered and more deeply marked. The blooms are shallow trumpets of pale lavender, spotted with orange-brown in the throat, and they are carried on 4 to 6-in (10 to 15-cm) stems in nodding clusters over the rough, hairy, leaf rosettes. *H. rhodopensis* is the easiest to find, and a gem you should cherish.

Cultivation Along with ramonda, haberlea is the plant that's alway recommended for that dingy, north-facing crevice. And quite rightl too, for that's where it is happiest – in a cool, shady spot. Lik ramonda, it enjoys being planted in a vertical fissure so that it rosette sits at right angles to the soil below. Here it finds the shar drainage much to its liking, but do make sure that the soil contain plenty of humus that can hold on to moisture in dry weather. A leafy, peaty, gritty mixture should surround the roots, but if yo find this difficult to hold in position in a vertical hole, ram aroun it a few pieces of turf, grass-side inwards, to give stability. Th grass will soon rot off, and the roots will bind the soil together. Th plant can also be grown in pans in the alpine house. Use a compos consisting of equal parts sieved leafmould, sharp grit and Joh Innes No. 2 potting compost. Wedge a few pieces of rock betwee the leaves and the compost. Water freely when necessary from spring to autumn and keep the compost just moist in winter. Stand the plant in a cool, shaded frame in summer, plunging it in moist sand or peat if possible. Repot after flowering when necessary.

Propagation In time the rosettes will grow large enough for you to divide in spring, but in the meantime you can take leaf cuttings. Pull off a mature, healthy leaf with its stalk, and insert it up to the base of the leaf blade in sandy compost inside a propagator. Taken in summer the cuttings can be potted in autumn and overwintered in a frame before being planted out the following year. Seeds can be sown when available in spring. Scatter them on the surface of peaty compost in pots, lightly cover with fine grit, and germinate in a frame. At no time let the compost dry out.

Hacquetia (Umbelliferae) Dondia

This bright spring novelty has tri-lobed leaves of shining green which are topped with a rosette of five green bracts centred with a boss of yellow flowers. It grows only about 4in (10cm) high, and exists in just one species: *H. epipactis* (syn. *Dondia epipactis*) (fig. 2.5). It comes from the European Alps, and is a must for your rock garden.

Cultivation An easy plant to grow in a shady spot on the rock garden, where it luxuriates in cool conditions and moist soil. Rather

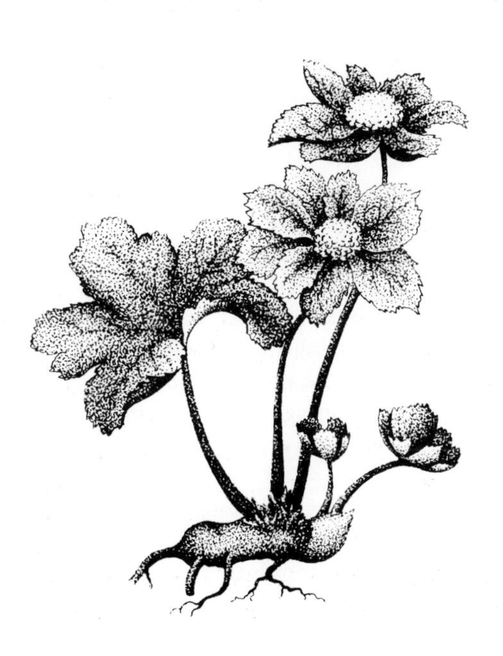

Figure 2.5
Hacquetia epipactis
Dondia

stiff earth seems to suit it best, and it hasn't yet turned up its toes in my chalky soil. Try not to disturb it once it's planted, for its thick roots travel deeply. Try it in a peat bed, where it will relish the rooting medium. It dies down in winter but soon emerges again.

Propagation Try to divide it in very early spring (February) if you can, but don't risk losing your plant by disturbing it too much. Seed is sometimes available, and can be sown in peaty seed compost in a frame in spring. Mature plants sometimes seed themselves, and the youngsters can be transplanted before they have had a chance to delve too deeply.

Helianthemum (Cistaceae) Rock Rose

Now you really can't do without rock roses whatever else you forgo. They make squat and spreading bushes that tumble over the ground, and from early to mid-summer they plaster their stem tips with ranks of glorious, wide-faced flowers centred with golden blobs of stamens. Yellow, crimson, pink, orange, white or bicoloured blooms may be borne, and the double varieties have the advantage of holding on to their petals for several days (those of the singles are shed by the evening of the day on which they opened). The narrow, evergreen leaves may be dark in colour or light grey. There are dozens of different varieties amply described in nurserymen's catalogues, so there's no need for me to elaborate, except to say that they are usually described as hybrids of *H. nummularium* and that they will all grow to around 6in (15cm) high and 3ft (90cm) across. But there is one miniature worth looking out for: *H. alpestre serpyllifolia* from central and southern Europe makes a low carpet of grey leaves topped with bright-yellow flowers. It's not so vigorous as the rest of its coarser relations, and can be planted where there's less space to spare.

Cultivation Bright, sunny banks are the best places to plant and view rock roses, but in any well-lit spot they'll do well if given room to spread their stems over rocks or rubble. Any ordinary, well-drained soil suits them. They'll tolerate drought once established and chalk right from the start. Cut them back a little with the garden shears after flowering to keep them in trim and prolong their life You might even get a second flush of flowers for your trouble.

Propagation Cuttings of shoot tips just 2in (5cm) long can be rooted in a propagator in summer. Seeds can be sown in spring in pots o seed compost, and germinated in a garden frame.

Helichrysum (Compositae) Everlasting Flower

Some of the 'alpine' helichrysums are a bit risky outdoors, but they're worth trying, and can certainly add foliar interest to an alpine house or frame. The weirdest is *H. coralloides* (New Zealand) – a coral-like plant with upright, branching stems composed of dark-green scales bedded on a white felt base. The overall effect is of a rather scaly but spineless cactus up to 9in (23cm) or so high. *H. selago* comes from the same country, but has thinner whipcord branches that splay outwards and downwards in an untidy manner. I don't find it half as attractive as *H. coralloides*, but you might. *H. milfordae* (syn. *H. marginatum*) (South Africa) is a different kettle of fish altogether. It makes masses of tiny, silver-haired rosettes like downy sempervivums but much smaller, and it covers the ground thickly. Red-budded, white flowers may appear on 2-in (5-cm) stalks in summer. *H. bellidioides* (New Zealand) creeps along the ground laying out its downy leaves and decorating them with 2-in (5-cm)-high, white flowers in summer. If you can try just one helichrysum outdoors this should be it.

Cultivation Warm, sheltered and sunny spots will grow the best helichrysums and also the ones that last longest. All four that I've mentioned have been grown successfully outdoors by lucky or skilful gardeners. Try planting them in scree beds, sunny crevices on the rock garden, gaps between paving stones, holes in tufa and well-drained soil in a sink or trough. They insist on sun and well-drained soil, but are otherwise not too difficult to please. The downy types will appreciate the protection afforded by a sheet of tilted glass in winter. Alternatively, grow the plants in pots in an alpine house or frame. They'll enjoy a mixture of equal parts John Innes No. 1 potting compost and sharp grit, and will need plenty of water during the growing season, but very little in winter. Keep the compost on the dry side at that time, and remove any dead leaves to prevent rotting. You might find this job rather difficult on *H. milfordae* (as I do), and you can risk leaving them on. If air circulation is good and water is not splashed around, the plant may even find its fading rosettes are good insulation against winter cold. Repot annually in spring.

Propagation The creeping and rosette-forming kinds can be divided in spring. The other two can be propagated from heel cuttings inserted in a propagator in late spring.

Hepatica (Ranunculaceae)

There are several species of *Hepatica* (formerly included in *Anemone*), but they are all so similar that I'll keep my comments to a minimum. The two you are most likely to encounter are *H. nobilis* (syn. *H. triloba*) (Northern Hemisphere) and *H. transsilvanica* (syn. *H. angulosa*) from eastern Europe. Both have dainty wood anemone flowers of lavender blue, centred with white-tipped stamens, but *H. transsilvanica* is altogether beefier. The leaves of both are three-lobed, slightly hairy and carried in a 4 to 6-in (10 to 15-cm)- high clump. They are a delightful sight in early spring.

Cultivation Hepaticas like a strange combination when it comes to soil. They must have a cool, moist earth, but prefer one that is chalky. If your soil is ordinary, mix in plenty of leafmould, together with a few limestone chippings or a dusting of ground limestone. If your soil is already chalky, stick to the leafmould. Find a shady spot – at the foot of a rock bank, wall or boulder – and plant this woodlander where it can be left undisturbed. The leaves tend to die down very late in winter, just before the new growth emerges.

Propagation Divide mature clumps just after flowering – if you've got the courage. Alternatively, gather seeds as soon as they are ripe, and sow them immediately in pots of seed compost placed in a garden frame.

Hieraceum (Compositae) Hawkweed

There are two or maybe three hawkweeds that I'd allow into my rock garden – I've quite enough wild ones in other parts of the plot! *H. aurantiacum* (Europe) has small, orange-red dandelions for flowers, and it spreads by means of creeping stems. It's not a plant to establish near slow-growing treasures, which it can swamp, but its colour is unusually rich and its common name of Grim the collier should endear it to Welsh gardeners. It will grow between 9 and 12in (23 and 30cm) high and blooms in summer. The other two plants are so similar that I'd be hard pushed to tell one from the other. Both *H. villosum* and *H. waldsteinii* (the first from central Europe and the second from Greece and the Balkans) make neat clumps of large, oval, white-felted leaves that are a bold contrast to more fussy plants. The foliage rises to not much more than 6in (15cm), but a hard, wiry flower stem will push up from among them to around 1ft (30cm) or more. It carries orange-yellow flowers that add nothing to the plant's appearance, and I usually cut them off. The whole plant dies down in winter.

Cultivation It looks superb in crevices between paving, retaining walls, rock banks and anywhere on the rock garden where it can bask in full sun. It will send its roots down happily into any well-drained soil – chalky or not – and is dead easy to please.

Propagation Divide mature clumps in spring, or sow seeds at the same time in pots of seed compost. Germinate them in a garden frame.

Hutchinsia (Cruciferae)

Some would say that *Hutchinsia* is an alpine weed, but that's being a bit hard. True enough, when it's finished flowering it does look a bit like our own hairy bittercress, but in bloom it's so flower-laden that it's worth growing. *H. alpina* (European Alps) is the kind most freely available. It grows just 3in (8cm) high, and is wreathed in small, white flowers in spring and early summer. The leaves are finely cut – rather like those of the aforementioned weed.

Cultivation Most at home in a sink garden or a hole bored in a lump of tufa, *Hutchinsia* needs no more than a sunny spot and any ordinary well-drained soil. It survives, too, in gaps between paving stones, and relishes scree conditions.

Propagation Divide clumps in spring. Sow seeds in spring in pots of compost or where the plants are to grow.

Hypericum (Guttiferae) St John's Wort

There's no place in any rock garden for the Rose of Sharon, *H. calycinum*, but there are some smaller treasures of great value for spring and summer flowers. The blooms of all my chosen species are bright, butter yellow with that hallmark of the hypericums, a central tuft of 'shaving-brush' stamens. *H. olympicum* (syn. *H. polyphyllum* of gardens) (Syria, Asia Minor and south-eastern Europe) should be your first choice. It makes 9-in (23-cm)-high, tumbling domes of foliage up to 2ft (60cm) or more across, and the small, grey-green, oval leaves always look smart. The flowers appear at the shoot tips, and are especially attractive in the variety 'Citrinum', which has pale primrose-yellow blooms. *H. coris* (southern Europe) is a charming miniature evergreen shrublet just 6in (15cm) high with linear leaves and large clusters of starry, yellow flowers carried at the stem tips in summer. It's not as hardy as *H. olympicum*, but even more tender is *H. cuneatum* (Syria and Asia Minor), which really needs alpine-house treatment. The foliage is blue-grey, the trailing stems burnished red and the flowers red in bud and yellow on opening – quite a display on a plant just 3in (8cm) high.

Cultivation In the open give your hypericums plenty of sun and a well-drained soil. They are happy on chalk, and look good cascading over rocks, pavements and walls. The two tender species are best planted in a scree bed if you want to risk them outdoors. With relatively mild winters they'll last a few years before you have to replace them. In pots the plants will be happy in a mixture of John Innes No. 1 potting compost and sharp grit with a few limestome chippings added. Water them freely when necessary in spring and summer, but keep them on the dry side in autumn and winter. Repot annually in spring.

Propagation Sow seeds in spring in pots of seed compost, and germinate in a frame. Take cuttings of shoot tips in summer, and root them in a propagator. Divide large clumps in spring or autumn.

I

Iberis (Cruciferae) Perennial Candytuft

Only *I. sempervirens* (southern Europe) is cultivated with any regularity on rock gardens, but it is really worth growing for its vast, spring flush of white, candytuft flowers held over a 9-in (23-cm)-high spreading mat of dark-green, leafy stems. Give it at least a yard (metre) to itself, or grow 'Little Gem', which is rather smaller. The best variety of stature is undoubtedly 'Snowflake', and all the varieties keep their leaves in winter.

Cultivation Tumbling down rocky banks, over walls and across paving, iberis looks very much at home. It really is quite unfussy about soil, thriving in stiff loam or sandy earth, and it doesn't mind chalk at all. A light clipping over after flowering will help to prevent the plant from falling open due to the considerable weight of its flowers.

Propagation Divide mature clumps in spring, or root cuttings of shoot tips in a propagator in summer.

J

Jasione (Campanulaceae) Sheep's Bit

These are interesting plants to grow if you have a spare corner. The blue flowerheads of *J. jankae* (Hungary) are carried on 9-in (23-cm) stems, and look like a cross between those of scabious and phyteuma. They rise up in summer from a tuft of hairy leaves. *J. perennis* is similar, but comes from western Europe and is not quite so showy.

Cultivation Bright sun and a well-drained soil are all these plants ask for. They'll put up with chalk, and can be squeezed into small pockets of soil on the rock garden, rock bank or raised bed.

Propagation Divide mature clumps in spring. Seeds can be sown in spring in pots of seed compost, and germinated in a frame.

Jeffersonia (Podophyllaceae) Twin Leaf

Although the alternative names of *J. dubia* (syn. *Plagiorhegma dubia*) and *J. diphylla* (syn. *Podophyllum diphyllum*) would put any gardener off growing them, they really are little beauties worth acquiring. *J. dubia* comes from north-east China (Manchuria to be precise), and sends up an orderly clump of the most beautiful wide-faced, lotus flowers of lilac-blue in early spring. The kidney-shaped leaves of grey-green tinged with purple follow quickly and shield the flowers as they fade. It's a plant no more than 4in (10cm) high, whereas *J. diphylla* (from North America) is up to 6in (15cm) or so. Here the grey-green leaves are definitely two-lobed and the flowers are white, again opening in early spring. These are mouth-watering miniatures.

Cultivation If you've a peat bed, grow your jeffersonias there. If not, find them a spot in a well-drained, lime-free soil that can be thoroughly enriched with leafmould and peat. They need shade and a cool root run to do really well.

Propagation Being clump-formers the jeffersonias can be divided in late summer or just before growth starts in spring. But if you've ever dug up a plant you'll know that below that modest tuft of foliage lurks a vast, fibrous root system that almost defies division. Rely instead on seeds, sowing them in spring or summer (as soon as ripe for preference) in pots of seed compost in a garden frame. Leave your clumps to multiply in size without being divided.

Kelseya (Rosaceae)

I've grown this curiosity in a pot for several years and sometimes wonder why. It makes a dense cushion of tiny, white-haired rosettes, and when it does flower (which is not often) its name of *K. uniflora* is not much of an understatement. The few flowers are pale pink in colour, and you'll have to be eagle-eyed to spot them. Still, kept in good condition, this native of the Western United States of

America is modestly pleasing to look at, and makes a good-sized cushion in time.

Cultivation Try it in a sink garden wedged among rocks, if you can keep it dry in winter. It is more reliably grown in pots of compost consisting of one part John Innes No. 1 potting compost to two parts sharp grit, with its crown sandwiched between two pieces of stone. Water it freely when necessary in summer, but more meanly in winter, when all moisture should be kept off the leaves and the compost just prevented from turning dusty. Repot in spring when necessary. Starved plants flower best.

Propagation I don't think you'll want many, but apparently the rosettes can be rooted in a propagator in early summer, and seeds can be sown in pots of sandy compost in spring, and germinated in a frame. One's enough for me.

L

Leontopodium (Compositae) Edelweiss

> Every season the misguided go dropping off precipices on which a few stray tufts have seeded down; not knowing that 200ft higher, in the soft alpine grass, they could be picking basins-full of blossoms in half an hour's gentle and octogenarian stroll before dinner.

Thus, in *The English Rock Garden*, Reginald Farrer kills the myth that edelweiss is a rare alpine of unattainable mountain ledges fit only for goats and gravel. But if the plant were known to be easily found, and if its other common name of flannel flower were adhered to, would folk still seek it out? Who knows? The fact is that *L. alpinum* (European Alps) is easy to find not only in the wild but also in nurserymen's catalogues, and it is planted by the hundred each year by fans of alpines and *The Sound of Music*. The bloom is a starry, grey-felted affair – glistening white it seems in the wild – and the foliage narrow and downy. It's a modest little plant that seems to have been prostituted by its press. If you fancy something a little different try *L. haplophylloides* (syn. *L. aloysiodorum*). It has many smaller heads of starry flowers and is scented of lemon – a fragrance that has resulted in its being called the lemonade Elweiss; sorry, lemon edelweiss.

Cultivation Edelweiss is only difficult to grow where drainage is slow, and the atmosphere is stagnant and moisture filled in winter. In any sunny patch of poor, sharply drained soil – chalky or not – it will thrive. Try it in a sink or trough garden or, better still, in a scree bed. If you live in an area of high rainfall, try protecting the tuft of leaves with a pane of glass in winter.

Propagation Divide mature clumps in spring, before flowering. Sow seeds in spring in pots of seed compost, and germinate in a garden frame. Keep the seedlings moist and shaded slightly in their early stages.

Leptospermum (Myrtaceae)

The varieties of *L. scoparium* (New Zealand) are a superb plant to have, but they are not reliably hardy. Grow them outdoors, and you'll find that even in the milder parts of the country they are hammered by winter frosts. However, for the alpine house or frame they are well suited, and bring considerable brightness to late spring and early summer. *L. s.* 'Nichollsii Nanum' is a compact treasure with the tiny, linear leaves that are typical of the species. They are burnished deep plum-bronze, and thickly clothe a bush just 10 or 12in (25 or 30cm) high. The flowers are apple-blossom shaped and deep pink. *L. humifusum* (Tasmania) *is* reliably hardy in Britain, and makes a wide-spreading, low thicket of growth which is studded with white flowers in summer.

Cultivation *L. humifusum* relishes a bright, sunny spot outdoors and a really poor, sharply drained soil free of chalk or lime. Grow it in a scree bed or as a part of a poverty-stricken alpine lawn. The forms of *L. scoparium* seem to be happy with me in a mixture of John Innes ericaceous compost and sharp grit in equal parts. Keep them well supplied with water in spring and summer when necessary, but on the dry side through winter and protected, where possible, from frost. Repot annually in spring, cutting out unwanted or dead shoots at the same time.

Propagation Cuttings of firm young shoots can be rooted in a propagator in summer. Seeds can be sown in spring in pots of seed compost, and germinated in a propagator.

Leucanthemum (Compositae)

There's just one member of this genus grown with any regularity and that's *L. hosmariense* (syn. *Chrysanthemum hosmariense*). It's nothing more than a dog-daisy, but a dog-daisy with the most finely cut, silver-grey, filigree foliage that makes a 6-in (15-cm)-high, 2-ft (60-cm)-wide mound. The white-petalled, yellow-centred flowers are produced off and on from spring to early winter. You can't ask for a better performance. There is one snag: the plant is not bone hardy in all parts of Britain. It has survived many winters in gardens in the southern part of the country, but I cannot speak of its track record in the north. Really severe frosts can cut it back, but it will often sprout again from lower down come the spring. Its Algerian origins explain its tenderness, but it's well worth a little experimentation in your part of the country.

Cultivation A soil that's sharply drained and not too rich, and a patch that's hit by every scrap of sun will suit the plant best. Try it on a scree bed, or a rock bank where it can tumble down a slope. In pots in an alpine house or frame it will grow in a mixture of equal parts John Innes No. 1 potting compost and sharp grit. Keep it well supplied with water during the growing season; much drier in winter when sogginess is not to its liking. If you decide to try it outdoors, a pane of glass will help to keep off rotting winter rains and a few degrees of frost.

Propagation Cuttings can be rooted in a propagator in summer.

Lewisia (Portulacaceae)

Afficionados with the lewisia bit between their teeth are driven to collect and grow to perfection as many species as they can. It's a task that's none too easy with some species, but others are reliable garden or alpine-house plants that anyone with a little know-how can tackle. There are around twenty species and multitudes of hybrids, and it is to these that the gardener who wants to brighten his 'rockery' will turn. *L. cotyledon* is the parent of most of them. Like all lewisias it hails from the USA, but it's about the easiest one to grow. It makes fleshy, evergreen rosettes of leaves from which spurt branched flower stalks in early summer. The starry blooms may be white, pink or salmon coloured. The many hybrids vary in their worth, but those bought from a good nurseryman will seldom disappoint, and may flower at any time from spring to late summer. *L. columbeana* is next in line when it comes to ease of culture and availability. It makes narrower-leaved rosettes than *L. cotyledon*, and has rather smaller flowers that are carried in larger numbers on multi-branched stalks. Its blooms may be anything from palest blush pink to rich rose, and it flowers in spring or early summer. *L. tweedyi* is a gem. It produces fewer flowers on smaller stems, but they are such wide-faced, starry blooms that they captivate all who set eyes on them. They are a glorious confection of salmon and pink, though other shades have occasionally appeared. The blooms open in spring and, sadly, the plant sometimes

-dest-sized rock bank in my garden just a few -hs after being planted up. There's still room for -ional plants which will eventually envelop the rocks -ing much of the gravel topdressing.

Bright colour and a luscious scent make daphnes some of the most desirable rock garden schrubs. *Daphne cneorum* 'Eximia' is, for my money, the best of the bunch.

a look at the flowers of *Dodecatheon pauciflorum* -ou 'll see why the plant is called shooting stars.

Alpines mix well with border plants creating bright edging in a sunny, well-drained bed at Beth Chatto's garden. *Alstroemeria* 'Ligtu Hybrids' are the pink flowers.

◀ If your rock garden lac
water, try a floral water
such as *Genista lydia*
which will be awash wit
bright yellow pea flower
in spring.

▶ Sink gardens are a boo
in small spaces and can b
crammed full of tiny
cushion formers. This on
at Harlow Car Gardens,
devoted mainly to
Saxifrages.

◀ Most thymes are easy
grow and overwinter, b
Thymus cilicicus can be
tricky out of doors. On
glance at its purplish-
pink bottle-brushes,
though, and you'll be
tempted to try it — sor
times with success.

▶ Rightly described as
'everyman's gentian',
Gentiana septemfida is
easy to grow and a real
treat when its blue
trumpets open in
summer.

The limestone rock garden at Harlow Car Gardens in North Yorkshire shows what can be done where there's the space to make a large and handsome rock feature. You'll need ample funds for the rock, too!

On sheer rock faces, down steps and over walls, the stems of *Euphorbia myrsinites* will tumble haphazardly to provide a subtle symphony of grey and greeny yellow in summer.

proves to be short lived. *L. rediviva*, unlike the other species mentioned, is deciduous. Its leaves are especially narrow, though still succulent, and they die away as the flower emerges in spring or early summer. Pink-and-white-flowered forms are offered.

Cultivation Most lewisias need lime-free soil that is rich in organic matter yet gritty and well drained, but they can sometimes be coaxed to grow on chalky ground that has been enriched with peat and leafmould. *L. cotyledon* and *L. columbiana* thrive in screes where a sufficiently rich mixture is present, and they look superb when growing out of a wall. Most lewisias relish being planted in vertical crevices so that their rosettes are at right angles to the ground – in this situation they are better equipped to shed the excess moisture that can lead to rotting of the rosettes. With the exception of *L. rediviva*, all the species I've mentioned do seem to flower best in a dappled or slightly shaded spot rather than one in full sun. Walls with northerly aspects can be confidently planted up with *L. cotyledon* hybrids to brighten their stony countenances. *L. tweedyi* is trickier. Try it in the upper course of a retaining wall, perhaps with a fraction more sun. One accomplished grower recommends removing any side rosettes in spring to improve air circulation around the central rosettes. The rosettes can be used in propagation and so are not wasted. After flowering the plant likes drier conditions at the roots for a while to afford it a welcome rest. *L. rediviva* likes a really good baking after flowering, and so is usually grown in pots in the alpine house where it can be kept quite dry from the end of flowering until early December when growth starts. All species can be grown in pots in a mixture of John Innes No. 2 compost and sharp grit in equal parts. Make sure with all species that the crown is held well above compost level by a layer of stone chippings – the same applies to plants grown on sloping ground (rather than vertical rock crevices) outdoors. Water them well as soon as the compost starts to dry out during the growing season, but take care not to overwater. Especial care must be taken in winter when the compost must be kept very much on the dry side (except for *L. rediviva* which is starting into growth). Repot annually after flowering. Young plants may need potting on twice a year. It's a rum genus, and Roy Elliott's monograph *Lewisias*, published by the Alpine Garden Society, makes fascinating reading, even if at times one despairs of ever cultivating the tricky members of the tribe.

Propagation Offsets can be rooted in sandy compost in a frame in spring. Seeds can be sown as soon as ripe in a peaty, sandy compost, though they may take up to a year to germinate. Leaf cuttings can be rooted in a mixture of peat and sand in a propagator in June.

Linaria (Scrophulariaceae) Toadflax

Closely related to (and jumbled with) *Cymbalaria*, *Linaria* is a fine plant for any rock garden, however small. *L. alpina* (European Alps) trails across the ground, rising only a few inches above the soil to show off its tiny snapdragon flowers of orange and violet purple. The thin leaves that radiate from the stems are a gentle blue grey, and the blooms are carried in summer. *L. maroccana* (North Africa) offers more of a rainbow of colours – its blooms being blue, white, yellow or red. It's a little larger at around 9in (23cm) and rather short lived – often lasting just one year. Not to worry – it produces plenty of seeds.

Cultivation On a neglected rock bank that I inherited just over a year ago, linarias sprouted from every inhospitable crevice, proving that they relish any hot, dry, poverty-stricken earth that basks in sunlight. Try them in screes, retaining walls, crevices between paving stones and alongside steps, where they will surprise you with a long season of flower. I've mentioned that *L. maroccana* is short lived, but I ought to warn you that *L. alpina* doesn't last long either. Mind you, it seeds itself so freely that you might not even notice – the youngsters will quickly replace the old-timers. Let loose in an alpine lawn, it will add its colour to the tapestry of mixed foliage.

Propagation Sow seeds in pots of seed compost in spring, and germinate in a frame. Alternatively, sow the seeds where the plants are to grow, also in spring. Longer-lived plants can be divided in spring.

Linnaea (Caprifoliaceae) Twin Flower

Some plants have a quiet grace that is captivating, and *L. borealis* is one of them. It's a woodlander from the Northern Hemisphere (including Scotland – though not much of it) which creeps along the ground, making dense mats of rounded, evergreen leaves. Above this verdant rug rise 3-in (8-cm) stems in June or July, each one carrying a pair of blush-pink flowers held back to back. Look inside each bellflower and you'll find deeper pink markings as though the plant were blushing at your attentions. How fanciful! The form most frequently offered by nurserymen is *L. b. americana* – the American version which is, as you might expect, rather larger in all its parts. It's also a bit easier to grow.

Cultivation If you've a peat bed, plant linnaea there. If not, find it a shady pocket on the rock garden which can be enriched with plenty of peat and leafmould to make a welcome home for its roots. It layers itself freely, but needs a hospitable mix to encourage such adventitious rooting. It needs lime-free soil. Make sure that the earth does not dry out in the heat of summer or the plant will suffer. Once the plant is established it might roam into places you would not consider conducive to its survival. One grower reports that the American form ventured over a sunny stretch of paving where it basked in the sun.

Propagation Divide mature mats in spring or autumn, replanting rooted portions.

Linum (Linaceae) Flax

These are valuable plants wherever summer colour seems to be lacking. *L. perenne alpinum* (syn. *L. alpinum*) (Europe) will grow anything from 6in (15cm) high to over 1ft (30cm), to display its bright-blue flowers in the sun. *L. narbonense* (southern Europe) is, if anything, bluer still, and a much taller plant at up to 18in (45cm). Even a small plantation is a breathtaking sight in summer. *L. flavum* (central and southern Europe) brings its own sunshine, for its blooms are a glorious yellow, carried on a more woody, branching plant than those of the previous species. It grows to about 1ft (30cm) high. The variety 'Compactum' is even more squat and useful where space is limited.

Cultivation Any ordinary, well-drained soil and full sunshine will grow good crops of flax. To bring height and colour to the rock garden, rock bed and soil-filled wall, they are a good choice. Snip out any damaged stems of *L. flavum* in spring.

Propagation Cuttings of young shoot tips can be rooted in a propagator in early summer. Divide clumps in spring. Sow seeds in spring in pots of seed compost, and germinate in a frame. Seeds can also be sown outdoors where they are to flower.

Lithospermum (Boraginaceae) Gromwell

We should really be calling this plant *Lithodora* now, but for the sake of understanding rather than botanical accuracy I stick to the old familiar name. A nurseryman told me recently that the last plant to be taken off his list would be *L. diffusum* 'Heavenly Blue', because it was such a good bread-and-butter line. The name must certainly inspire gardeners to buy, but the plant has now really been superseded by the variety 'Grace Ward', which possesses more vigour. Both plants make sprawling mats clad in bristly, dark, evergreen leaves, and from the tips of which erupt sky-blue flowers all through the summer. *L. oleifolium* (Pyrenees) is not so striking in colour, but it's still a good plant. It grows to around 4 or 6in (10 or 15cm), and from among its tufts of silvery-haired leaves spring nodding clusters of pinkish-mauve buds that turn blue as they open. It's just as generous with its flowers as *L. diffusum*, blooming right through summer.

Cultivation Find your lithospermums a spot in full sun where they can cast their wandering stems over rocks, walls or steps. *L. diffusum* insists on a lime-free soil that's sharply drained, and it enjoys a bit of peat or leafmould enrichment (plus grit if the soil is stiff). *L. oleifolium* is the species to choose if you garden on chalk (unless you can make *L. diffusum* feel at home in a sink of gritty, acidic compost as I have done). Trim off any dead stems in spring, and try to have a few young plants at hand to replace old specimens which have become straggly or which have died of exhaustion.

Propagation From time to time you might hear complaints that growers are having difficulty in rooting cuttings of lithospermum. This does happen, usually as a result of over-propagation for many years, which drains the plant of its vigour. It's a fault only with *L. diffusum* – usually in its variety 'Heavenly Blue'. However, if 2-in (5-cm) long stem cuttings of non-flowering shoot tips are taken in late July, they will usually root in good numbers in a shaded garden frame containing a mixture of equal parts peat and lime-free sand which is kept constantly moist (not soggy). One grower recommends that the lower leaves are not removed from the cuttings as they seem to aid rooting. Remove and pot up the cuttings as soon as new growth is observed, taking great care not to damage the roots. Layered shoots of *L. oleifolium* can be removed and transplanted in autumn or spring. The plant naturally makes underground runners.

Lychnis (Caryophyllaceae) Campion

Alpine relatives of the red campion, one or two species of *Lychnis* are worth growing. *L. alpina* (syn. *Viscaria alpina*) (Europe, includ ing Britain – in the Lake District and Scotland) is the smalles species, seldom more than 4 or 6in (10 or 15cm) high. From rosette of green, red-tinged leaves arise reddened stems that hold alof clusters of dull-pink flowers. The blooms open in early summer *L. flos-jovis* (European Alps) is altogether more interesting with it tufts of leaves thickly coated in silvery down. The foot-high (30-cm) flower stems rise up in summer to display rich cerise-pink flowers at the top. There's a more compact form called 'Nana', and 'Hort's Variety' has blooms of a more easily placed pink.

Cultivation Any ordinary soil that's sharply drained will keep *Lychnis* happy, and chalk is no deterrent to their growth. Full sun is essential – try the plants on sunny slopes in the rock garden, in large or small pockets of earth between paving stones, and on top of retaining walls.

Propagation Divide mature clumps in spring. Sow seeds in spring in pots of seed compost, and germinate in a garden frame. Some self-seeding may occur, which indicates that the plants are easy to sow *in situ*. If this is a problem, snip off the flowerheads at crown level when the blooms have faded.

Lysimachia (Primulaceae) Creeping Jenny

I've only one *Lysimachia* to recommend to you and that's *L. nummularia* 'Aurea', the yellow-leaved version of the creeping Jenny native to Europe, including Britain. But what a plant it is. Spreading flat along the ground it clothes its stems in rounded, acid-yellow leaves that brighten any shady nook. Its yellow, cup-shaped star flowers open in summer.

Cultivation A moist, peaty, leafy loam will keep this squat but fleshy plant healthy, and a position in gentle shade will ensure that its vigour and leaf colour are kept to maximum strength. It doesn't mind chalk, provided that the soil is not prone to drought. Find a flat or sloping stretch of soil on the north side of the wall or boulder, and try a patch of creeping Jenny there.

Propagation This is dead easy. Remove rooted runners and transplant them at almost any time of year. Only drought will prevent them establishing quickly.

Mazus (Scrophulariaceae)

These little creepers deserve to be more popular. *M. reptans* (Himalaya) is the most widely offered species, and makes flat, creeping mats of green leaves which are decorated with tiny, lobelia-like flowers of pale violet spotted with orange. The blooms appear at any time in late spring and summer. *M. pumilio* (New Zealand) is similar, but lacks the creeping habit of the former, making neater mats. I grow a white form which is not exactly spectacular, but it earns its keep.

Cultivation This is not a difficult genus to please, and you'll be able to grow the plants in any reasonable soil in full sun. They will do well in rock beds, on banks and as part of alpine lawns, and *M. pumilio* is a natural candidate for trough and sink gardens. Old plants may become straggly, and are then best replaced with youngsters.

Propagation Divide clumps in spring.

Meconopsis (Papaveraceae) Blue Poppy

Woolly mammoths they may be, but I can't omit *Meconopsis* from my list – so many alpine growers find them a space at the back of the rock garden or even in part of a border. They're true alpines, mostly native to those fascinating provinces China, India, Kashmir, Nepal, Tibet and Burma, and they've inspired gardeners and plant collectors alike since the earliest days of eastern expeditions. There are many species, all but a few of which are difficult to obtain from nurserymen, though a wider variety can be bought in seed form, and that's how I advise you to start. *M. betonicifolia* (syn. *M. baileyi*) (Tibet, Upper Burma, Yunnan) is the fabled 'Blue Poppy' and a plant that proves reasonably perennial, provided it is not allowed to flower until it has built up several rosettes (pinch out earlier flower stems). Many species are biennial or monocarpic (dying after flowering), and so whatever you grow, frequent re-sowing is advisable. The blue poppy will grow to 3ft (90cm) or more, and in its best forms has papery blooms of shining azure. *M. grandis* (Sikkim, Nepal, Bhutan, Tibet) is similar in colour, but flowers earlier and has leaves with fewer lobes. It too is reasonably reliably perennial, and its form 'GS 600' is the most vigorous and worth seeking out. *M. cambrica* (Europe, including Britain) is the Welsh poppy – a poor relation well worth mentioning. It grows to about 1½ft (45cm), has mid-green, finely cut leaves that are softly hairy, and produces plenty of yellow poppies in late spring and early summer. The trouble is that it's a bit too easy to grow – it seeds itself everywhere in my garden; and though I welcome its early blooms, I wish it would restrain itself from spreading! It is definitely perennial! *M. napaulensis* (Nepal and South West China) is the satin poppy, and offers perhaps the widest colour range of any meconopsis. It grows to between 4 and 6ft high (1m 25 and 1m 85cm), is profusely hairy (usually with reddish bristles), bears blooms of deep red, purple, blue, pink, white, apricot or yellow in tall spires, and dies after flowering.

Cultivation Some folk have succeeded with *Meconopsis* on chalky ground, but usually the only one happy under such circumstances is *M. cambrica*. The rest enjoy a peaty, leafy soil in partial or dappled shade, though they will take full sun for part of the day provided that the soil stays moist (this does not mean that they will put up with poor drainage). Grow your plants in a peat bed if you have one, or make a large pocket of soil that will suit them in a shady part of the rock garden. The monocarpic kinds like *M. napaulensis* will take several years to reach flowering size, but during that time you can admire their highly decorative leaves. Pull up and discard dead plants, making arrangements to relieve them of their seeds before you do so.

Propagation Save seeds from your best plants, and sow these as soon as they are ripe. I find that they germinate quite easily in pots of peat-based seed compost placed in a garden frame in spring. Make sure that the compost does not dry out at any time. The flower colour of plants will vary in its intensity – sometimes a batch of *M. betonicifolia* seedlings will produce murky-flowered specimens, and these should be rogued out immediately. Keep selecting the best plants; collect seed only from them, and you will produce a more reliable race of bright-flowered aristocrats. Seeds bought from seedsmen have no such guarantee, and you can only hope for the best (and buy from a reputable source).

Mentha (Labiatae) Mint

For goodness sake don't let ordinary mint loose anywhere near your alpines or you'll never see them again. Two species that you can introduce though are *M. pulegium* (Northern Hemisphere), and *M. requienii* (Corsica and Sardinia). The first is pennyroyal – a plant that makes a dense green carpet of shiny, peppermint-scented leaves on top of which lilac-purple flower clusters may appear in late summer and autumn. *M. requienii* is the Corsican mint and an even smaller-leaved carpeter, reminiscent of 'Mind Your Own Business' (soleirolia or helxine), with minute purple flowers in summer. It's equally minty in scent.

Cultivation Both plants make superb carpeters for alpine lawns, gaps in patios and paths and for edging the rock garden. The Corsican mint is the fussier of the two, preferring a soil that is never likely to become dry and a spot that is not too sun baked. Pennyroyal can take as much sun and heat as is available – it thrived in very sandy soil among paving stones in my last garden – no doubt relishing its cool root run beneath the slabs. Both species are happy to grow in chalky soil.

Propagation Divide mature plants in spring or autumn.

Mertensia (Boraginaceae) Virginian Cowslip

Strictly, I suppose, this is more a plant for the flower border, but I've been so entranced with its beauty as a rock garden plant that I shall continue to think of it as such. *M. virginica* (North America) makes clumps of elegant arching stems clothed in soft, grey-green leaves. The stem tips form croziers from which hang tubular blooms – pink in the bud and sky blue when they open. It grows to around 1½ft (45cm), and is a graceful treasure.

Cultivation This is a plant for gentle shade and a soil rich in leaf-mould, which will help to prevent unwanted drying of the roots. The plant flowers in spring and its foliage dies away shortly afterwards – a factor which might be useful where space is in short supply and plants have to earn their keep. It is most at home in a peat garden or below a north-facing rock. Other, smaller species are equally fond of positions that are not sun baked.

Propagation Divide mature clumps in autumn, or sow seeds in pots of seed compost as soon as ripe, germinating them in a garden frame.

Mimulus (Scrophulariaceae) Monkey Flower, Musk

No summer flower could be brighter than *Mimulus*, not only in its ground colours of red, orange, yellow or white, but also in its spotting and blotching in a contrasting yet equally vivid hue. There are many hybrids with which you must take your chance, but not many will disappoint you. Most are derived from *M. cupreus* (Chile), and old favourites such as 'Whitecroft Scarlet' continue to delight new generations of gardeners. *M.* X *burnettii* is a brilliant-orange hybrid, yellow in the throat. New and ever more glorious varieties seem to appear each year. Other species are sometimes offered, and cultivation is the same for most of them as for the hybrids.

Cultivation Moisture at the roots is essential for *Mimulus*; indeed it often grows in boggy areas near garden pools. Make sure it has a good, stiff soil, rich in organic matter, that is never likely to dry out. That assured, it can be grown in dappled shade or full sun. It is tolerant of chalk.

Propagation Divide mature plants in spring. Take cuttings of shoot tips and root in a propagator in summer. Sow seeds in spring in pots of peaty seed compost, and germinate in a frame.

Minuartia (Caryophyllaceae) Sandwort

Tiny plants related to *Arenaria*, not many of these are found in nurserymen's catalogues. *M. verna* (Europe, including Britain) is the spring sandwort, and it can be located without too much difficulty. It presents few problems in cultivation, making needle-leaved domes of fresh-green leaves which send out tiny, white flowers on fusewire-thin stems in spring. *M. stellata* (syn. *M. parnassica*) (eastern Europe) is smaller and tighter in habit, and its tiny, white flowers sit direct on the foliage, also in spring.

Cultivation Good plants for sun or dappled shade in sink and trough gardens where they will enjoy a gritty, leafy soil. They are happy too in screes, provided they do not become too dry at the roots, and in pockets of soil between paving stones.

Propagation Divide clumps after flowering. Seeds can be sown in spring in pots of seed compost, and germinated in a frame.

Morisia (Cruciferae)

A unique little crucifer, *M. monantha* (syn. *M. hypogaea*) makes rosettes of dark, glossy-green leaves with pronounced saw edges. The rosettes hug the ground, and from their centres in spring emerge stemless, bright-yellow flowers. The plant is native to Corsica and Sardinia, and although it's no great beauty, it's a curiosity I would not want to be without.

Cultivation *Morisia* is mostly grown in pots in the alpine house, where it seems happy in a mixture of equal parts sharp grit and John Innes No. 1 potting compost. Water it freely when necessary in spring and summer, more cautiously in winter, keeping the compost very much on the dry side. Pull away any faded leaves with tweezers to prevent rot setting in. It makes long, thick roots, and is best grown in fairly deep pots. Repot annually after flowering. If you want to grow the plant outdoors, try it in a sunny scree bed or sink garden where drainage is very sharp.

Propagation Seeds can be sown in pots of seed compost in spring, germinating them in a garden frame. But the plant is often propagated by root cuttings, which can be taken from mature plants in June. Knock the plant free of compost, and remove suitable roots at least bootlace-thick, cutting them into sections 1in (25mm) long and *keeping them the right way up*. Insert these in a tray or pan which has been filled to the half-way mark with John Innes No. 1 potting compost and then to the rim with sharp sand. Pot up the plants individually when they have produced new rosettes.

Myosotis (Boraginaceae) Forget-me-not

The alpine forget-me-not, *M. alpestris*, is a rare British native, occurring only in Upper Teesdale and Perthshire, but it is widely distributed elsewhere in Europe, and brightens any rock garden in late spring or early summer with its sky-blue flowers lined with white and centred with a yellow eye. It only grows to around 3in (8cm) and is not that long lived, but it always supplies some seeds to carry on the line. *M. rupicola* (European Alps) is similar (and closely related) but even more dwarf at up to 2in (5cm).

Cultivation *M. alpestris* will grow in any well-drained, gritty soil on a rock garden, rock bed or bank, and in sink or trough gardens, but is easily at its best in a scree. It needs full sun, as does *M. rupicola*, which also grows best if given a really lean diet in compost consisting mainly of stone chippings and sand. Try it in a scree bed or a trough filled with scree mixture or in a hole bored in a piece of tufa. The two are occasionally grown in pots – *M. alpestris* will thrive in a mixture of equal parts John Innes No. 1 potting compost and sharp grit, while *M. rupicola* needs something much poorer: two parts of sharp grit, and one each of silver sand and sieved leafmould – a witches' brew, but one that will keep it happy! Repot when necessary after flowering. Water the plants freely when they are thirsty in spring and summer, but keep them only just moist in winter.

Propagation Sow seeds in pots of seed compost in summer, and germinate them outdoors. Cuttings of young, green shoots can be rooted in a frame in early summer, and plants can also be divided in spring.

N

Nierembergia (Solanaceae) Cup Flower

The large, white, upturned goblets of *N. repens* (syn. *N. rivularis*) are a glorious sight from early to mid-summer, erupting almost straight from the ground. The foliage is small and dark green, sitting close to the earth, and allowing the blooms to show off without competition. The plant comes from the Argentine and Chile, often growing in boggy ground, but manages to grow well in British gardens without its native mud.

Cultivation Find for the cup flower a spot in full sun in a gritty, well-drained soil that sheds excess moisture in winter yet does not dry out in summer. A scree bed will suit it fine, provided summer drought can be avoided. The plant can also be grown in the alpine house in wide pans containing a mixture of John Innes No. 1 potting compost and sharp grit. Water freely when necessary in spring and summer, much more carefully in winter, when the compost should be kept on the dry side. Repot in spring when necessary.

Propagation Divide mature plants in early spring (the plant produces masses of underground stems).

O

Oenothera (Onagraceae) Evening Primrose

If you long for a bright-yellow flower to take up the torch after *Alyssum saxatile*, then choose an evening primrose. *O. missouriensis* (USA) is just as greedy when it comes to space, but even more spectacular in flower. Its massive, wide-faced, sunshine-yellow blooms stud its upturned, spreading stems for weeks on end from early to late summer. As its common name implies, it's a flower that stays open in the evening – a delight to the worker who usually returns home to find that the rock roses have shed their petals and *Convolvulus mauritanicus* has closed up! *O. acaulis* (Chile) is the species to choose where space is short. It makes ground-hugging rosettes from which the large, yellow flowers emerge, fading to pink as they age. Several nurseries offer *O. glaber*, a species of uncertain origin and uncertain nomenclature, which is worth having on account of its bronze foliage, red buds and yellow flowers that open atop foot-high (30-cm) stems.

Cultivation Bright sun is essential, and any good, well-drained soil is usually tolerated, even if it's chalky. Plant them on sunny slopes on rock banks, in drifts on the rock garden, at the top of retaining walls and in pockets of earth among paving stones, where they will enjoy the sun up top and the cool root run down below. *O. acaulis* (like some other members of the genus) is often short lived, and may have to be regularly renewed from seed.

Propagation Large clumps can be divided in early spring. Seeds can be sown in spring in pots of seed compost, and germinated in a garden frame.

Omphalodes (Boraginaceae) Navelwort

Why is it that sky-blue flowers are always so captivating? The dwarf forget-me-nots, the ungrowable *Eritrichium nanum* (not included in this book for that reason, coupled with its unavailability) and some gentians have such charisma that they present an irresistible challenge. So it is with *Omphalodes*. *O. verna* (European Alps), known as blue-eyed Mary, makes 6-in (15-cm)-high plants with heart-shaped green leaves and loose sprays of rich blue flowers in spring – before any of the others have opened. *O. cappadocica* (southern Europe) is similar but rather larger in leaf and flower, blooming just slightly later. The jewel in the crown is *O. luciliae* – one of those alpines that stirs the pit of your stomach, it's so stunning. It comes from Greece and Asia Minor, and sends up a tuft of blue-grey leaves in spring. These are followed by long sprays of soft-blue forget-me-not flowers which last from early summer right through until autumn.

Cultivation The first two species I've mentioned are happiest in gentle shade, dappled by a nearby shrub or overshadowed by a rock face. Any well-drained soil that doesn't dry out in summer will keep them in good health, but helpings of peat and leafmould will make the earth most hospitable. *O. luciliae* prefers a much sharper medium, and a sunny scree is much to its liking – especially a limestone scree. Failing this, grow it in a stony, gritty pocket of limestone chippings on a sunny rock garden – it relishes good light, unlike its relations. A sink garden filled with a limestone scree mix will also grow good plants. In the alpine house or frame the plant can be grown in pots consisting of equal parts John Innes No. 1 potting compost and limestone chippings. Water freely when dry during the growing season, and keep just moist through the winter. Repot annually in spring.

Propagation Plants can be divided in spring, though you may be reluctant to dig up and separate a plant of *O. luciliae*. Raise this species from seeds sown in pots of seed compost as soon as it is ripe, and germinated in a garden frame. Self-sown seedlings sometimes appear, and these can be planted or potted up.

Ononis (Leguminosae) Rest Harrow

There's just one rest harrow offered with any frequency and that's *O. cenisia* (syn. *O. cristata*) from the Alps of Europe and North Africa. It's a tiny, sprawling plant with three-lobed leaves and pea flowers of rich pink carried on 2-in (5-cm) stems in spring and early summer.

Cultivation This miniature treasure is for a sink garden where it can be closely admired, or for a scree, where the gritty, stony soil suits it perfectly. Sun and brilliant drainage are its requirements.

Propagation The roots are deep and delving, making division impossible and transplanting very difficult. Raise new plants from seed sown in spring in pots of seed compost, and germinated in a frame.

Onosma (Boraginaceae) Golden Drop

Not always available in quantity, *O. tauricum* (south-eastern Europe) is well worth acquiring. It makes a plant around 9in (23cm) high, the leaves and stems of which are densely clothed in silvery hairs. The flowers – tubular bells of rich yellow – are carried at the tips of the shepherd's-crook stems in summer. *O. albo-roseum* might be easier to find, and has dainty, dripping bells of pale yellow and pink. Again the leaves are downy.

Cultivation In full sun and in gritty, stony soil on a rock garden or bank, *Onosma* will do well. It loves warmth and good drainage, and is a natural candidate for the scree bed. It looks good, too, when grown in the crevices of walls and it doesn't mind chalk.

Propagation Cuttings of young shoot tips can be rooted in equal parts sand and peat in a frame in early summer. Seeds can be sown in spring (or as soon as ripe if collected) in pots of seed compost. Germinate them in a frame.

Origanum (Labiatae) Dittany

No, this is not marjoram. There are at least two species widely available that make really good rock plants. *O. laevigatum* (Asia Minor) makes an upright, open bush about 1ft (30cm) tall. Its stems are deep maroon and its small, oval leaves grey green – a fine complement to the small, purplish-crimson flowers that are carried in late summer and autumn. Its time of flowering alone makes it worth growing, and its smart habit seals the deal. *O. tournefortii* is a Grecian dittany with stiff stems that carry rounded leaves wrapped in white wool. Summer heads of pink, hop-like bracts are a lovely sight alongside the leaves.

Cultivation Any sunny spot in a rock garden, bed or bank suits origanums, and a well-drained soil, chalky or not, is what they need around their roots. *O. tournefortii* can be grown in scree beds or pots in an alpine house or frame. A mixture of equal parts John Innes No. 1 potting compost and sharp grit suits it well. Water it freely in summer when necessary, but much more cannily in winter when the compost can be kept *almost* dry. Repot annually in early spring. Outdoors the plant will enjoy a bit of protection from winter wet – sit a pane of glass over it if you can. Cut out old, unwanted stems in spring, and any dead growths or faded flower stems as soon as they are seen.

Propagation Cuttings of young shoot tips can be rooted in a propagator in summer.

Othonnopsis (Compositae)

This is a North African native which proves surprisingly hardy in sheltered British gardens. *O. cheirifolia* is an evergreen, spreading plant up to 9 or 12in (23 or 30cm) high, and possesses remarkable, fleshy, grey-green leaves that are paddle-shaped and held in fan-like clusters. Over this sprawling mass of succulent leaves appear in summer glorious, broad-petalled daisies of bright yellow. It's a distinctive plant that gardeners with an eye for form and foliage should take under their wing.

Cultivation *Othonnopsis* needs shelter and sun to make it feel at home. Site it where it can tumble over a wall or a flat boulder. A gritty, well-drained soil will keep it happy, and any dead shoots should be cut out as soon as seen. It can also be grown in pots in an alpine house or frame. Use a compost consisting of equal parts sharp grit and John Innes No. 2 potting compost, and apply plenty of water when required through the growing season, much less in winter, keeping the compost barely moist. Protect the plant from cold winds and severe frosts. Repot annually in spring.

Propagation Shoot-tip cuttings can be rooted in a propagator in early summer.

Ourisia (Scrophulariaceae)

Of these engaging woodland plants, *O. coccinea* (said to be a synonym of *O. elegans*) is the most brightly coloured species, with nodding scarlet flowers at the top of 9-in (23-cm) stems between early and late summer. It comes from the Chilean Andes, while the other commonly cultivated species hails from New Zealand. *O. macrophylla* makes a thick clump of rough, heart-shaped leaves, over which are held rounded clusters of white flowers in July. At a foot (30cm) high it is rather larger than the previous species and altogether more robust.

Cultivation Both are most at home in a peat bed where they will luxuriate in gentle shade and a deep, humus-rich medium. Failing this, give them a peat and leafmould-enriched pocket at the foot of a north-facing boulder on the rock garden. *O. macrophylla* is quite a spreader, so needs a bit of room to breathe. If a stream runs through the rock garden they are a natural choice for the moist ground on its banks.

Propagation Divide mature clumps in spring. Seeds can be sown as soon as ripe in pots of seed compost, and germinated in a frame.

Oxalis (Oxalidaceae) Wood Sorrel

I've spent hours eradicating the pernicious *O. corniculata* from a rock bed, and so have to shake myself into remembering that there are species of this pretty genus worth including. Indeed, the rock garden would be a duller place without two in particular. *O. adenophylla* (Chile and Argentina) grows from a fat tuber – rather like a cross between a cyclamen and a crocus. Its leaves are works of art in themselves; folded with a craftsmanship equal to origami, they are blue-grey in colour and a superb backcloth to the pale pink, deeper-veined flowers. The leaves emerge in early spring, and the flowers a few weeks later. *O. enneaphylla* comes from the Falkland Islands, and is more of a spreader (though a very restrained one compared with my enemy the weed). Again, the leaves are a cool grey-green and the almond-scented flowers pink or white. Instead of a fat tuber, this species has a string of smaller ones beneath the soil. There's a lovely story told by the late Clarence Elliott, who presented Lily Langtry with a pink bloom of *O. enneaphylla rosea* at the Chelsea Flower Show, explaining that the Falkland Islanders used it to make a cooling drink. Miss Langtry pushed the flower through her veil and ate it. 'Whether she felt embarrassed at the thought of carrying so ridiculous a small flower round the show, or whether she was just out for a new gastronomic sensation I have never been able to decide', he wrote.

Cultivation Find both these plants a sunny or partially shady crevice on the rock garden where the soil has been enriched with grit and leafmould. *O. adenophylla* tends to work its way to the surface, so top-dress the tuber with a little leafy soil each spring. Both species can be grown in pots in the alpine house in pots containing a mixture of equal parts John Innes No. 1 potting compost, sharp grit and sieved leafmould. Keep the compost gently moist in summer, much drier (but not baked) in winter. Repot annually in January, setting the tubers about an inch (25mm) below the surface of the compost.

Propagation This is by division of tubers in spring just before growth starts.

P

Papaver (Papaveraceae) Alpine Poppy

The alpine poppies are a jolly bunch of miniatures; even if they are not long lived, at least they usually supply plenty of seed to carry on the good work. *P. alpinum* (European Alps) can be had from every nursery and almost every market stall. Its 2-in (5-cm)-wide poppy flowers may be rosy pink, white, yellow or orange, and appear over the finely cut leaves from early summer onwards. It reaches a height of no more than 6in (15cm). Few other species are offered, but if you ever come across *P. miyabeanum* (syn. *P. fauriei*), invest a pound or two. Its leaves are tufts of blue-grey fern, intricately cut and quite hairy, and its flowers are open bells of acid yellow on stems around 4in (10cm) high. It's Japanese, short lived, and my plant took its leave in the winter of 1981/2 in spite of being in a frame. It wasn't until the following June that I noticed a fine crop of seedlings in pots of dormant irises that had sat nearby. I now have a legacy of more than a dozen plants from the original.

Cultivation Bright sun and a gritty soil – preferably a scree bed – will suit these poppies down to the ground. Try them in gritty pockets among paving stones and on top of retaining walls. Left to their own devices they'll seed themselves happily and continue to thrive. They love chalk, and slugs love *P. miyabeanum*.

Propagation This is by seed sown *in situ* or in pots in a frame in spring.

Paraquilegia (Ranunculaceae)

This magical plant has acquired an intimidating reputation for being difficult, and when I first acquired it I felt sure that death would be imminent. It thrived. It's not that I'm a highly skilled grower, simply that if its modest requirements are met, it grows lustily. The only species freely available is *P. grandiflora* (syn. *P. anemonoides*) from Kashmir (the very place adds a certain mystique). During the winter there's nothing to look at other than a resting crown clad in a few fusewire-thin stem bases that you'll not dare shave off if you've any sense. In spring it starts to erupt into a dome of the finest cut silver-green filigree foliage you've ever seen – no more than 2in (5cm) high. Then, very slowly, flower buds start to appear, gradually swelling on the ends of their stems until they are held just clear of the foliage. They expand into the most beautiful open bells of lavender blue, centred with a blob of golden stamens. As they fade the foliage remains, until the autumn, when it turns dusky brown and dies. What a plant!

Cultivation Try it in a scree bed – sunny and sheltered – if you dare, but I grow mine in a pot. Not only is it easier to manage, but you can observe, gloat and appreciate it at close quarters much more easily that way. Mine is potted in the usual mixture – equal parts John Innes No. 1 potting compost and sharp grit, plus a scattering of limestone chippings (or a dusting of ground limestone). During spring and summer it is watered whenever it is dryish, but through the winter in the garden frame it stays much drier – a trickle of water being applied only when the compost looks dusty. Chop off the leaves to within ½in (13mm) of the crown as soon as they have died off, or the rot may spread back. Repot the plant when necessary after flowering. That really is all there is to it. The plant can take any amount of cold, provided that it is not soggy at the roots or the crown.

Propagation Although my plant forms seed pods, there's never anything in them, and as seed seems to be the only method of increase that makes things a bit slow! When seed is set it should be sown as soon as ripe in July or August – in April if you receive seeds during the winter. Sow them in a sandy seed compost, and germinate them in a frame.

Parochetus (Leguminosae) Shamrock Pea

Only one species exists, and that's *P. communis* from East Africa and the Himalaya. It's a trailing vetch-like plant with green, clover-shaped leaves and lots of lovely blue pea flowers in late summer and autumn – even on into winter if grown in an alpine house.

Cultivation Grown outdoors the plant is not totally hardy, and is likely to be killed back to ground level by severe frosts. But be patient and persistent, and new growths should emerge in spring. The plant enjoys best a soil that is rather moist – never likely to dry out – but faulty drainage is not something you should strive for. A sunny spot at the foot of a rock garden will naturally be more moisture retentive than a site at the top, and it will offer more shelter, which is also welcome. Try the plant near a stream if you have one. It doesn't mind chalk. In pans in an alpine house or frame the plant will bloom on well into winter. It's happy in a mixture of equal parts John Innes No. 1 potting compost and sharp grit, and can be repotted in spring when necessary. Water it freely when necessary in spring and summer, and keep it barely moist in winter. Prevent the plant from being badly frosted.

Propagation Cuttings of shoot tips can be rooted in a propagator or frame in summer. Naturally rooted stems can be removed and transplanted in late summer, and a few potted up and used as insurance against the plants outside being lost. Seeds can be sown in pots of compost in spring, and germinated in a garden frame.

Penstemon (Scrophulariaceae) Beard Tongue

Penstemons are a delight on the rock garden, but a nightmare when it comes to nomenclature. Nobody seems to know what's what, with the result that plants sold by one nurseryman as this, could be sold by another as that. Still, you'll not go far wrong with the following species, all of which come from north-western USA, and all of which are evergreen. *P. newberryi* makes a low, twiggy bush of dark-green leaves that are smothered in summer with foxglove-like bells of rich rose red. *P. menziesii* is similar, but has violet-purple flowers. It's also a little smaller at around 6in (15cm) – *P. newberryi* will make 9in (23cm). *P. scouleri* will eventually make 10in (25cm) or 1ft (30cm), and over its mound of leaves carries dense spires of lilac-purple flowers, or white ones in the form 'Alba'. *P. rupicola* is a real cracker. Its mat of green leaves is thickly set with rose-red bell flowers in summer, and it grows no more than 3in (8cm) high. Finally, a mention of *P. pinifolius*, which is a needle-leaved plant from 4 to 9in (10 to 23cm) high (though much more in the wild), and carries tubular flowers of burning scarlet in summer.

Cultivation Most plants with myriads of bright flowers love a spot in full sun, and penstemons are no exception. But they are an exception to the rule that alpines prefer a poor soil to a rich one, for they always do best when given a good diet. I don't mean that the ground they are to occupy should be heavily manured, just that it should be well enriched with organic matter and a sprinkling of blood, bone and fishmeal. What's more, make sure that it won't bake in summer. Many penstemons have been lost because they've been starved of food and drink in some narrow crevice or baking scree. Good soil and full sun should see your plants well set up for a long, bright life. Plant them where their mats can fall over rock faces, or on top of retaining walls so that they can brighten the brickwork, but if there's a chance of drought, be on hand with the hosepipe. The plants will usually put up with chalky soil provided it's well laced with humus. In pots penstemons prolong the display in any alpine house, blooming well into summer. Pot them up in a mixture of equal parts John

Innes No. 2 potting compost and sharp grit, repotting them every spring. Water freely when necessary from spring to late summer, but then allow the compost to remain on the dry side right through winter.

Propagation Cuttings of shoot tips can be rooted in a propagator in summer.

Phlox (Polemoniaceae)

Gaze at a well-grown carpet of alpine phlox in late spring and I doubt if you'll see a single leaf, so thickly does it pack its flowers. *P. subulata* (north-eastern USA) has provided most of the popular garden varieties which make thick, evergreen, needle-leaved rugs smothered with blooms of many shades, from deep violet through crimson to pink, salmon, pale blue and white. Favourites include 'Oakington Blue Eyes', with blooms of a glorious sky blue, and 'Temiscaming' with rose-pink blooms in equal profusion. The plant sold as *P. amoena* 'Variegata' is worth having, for as well as carrying plenty of pink flowers, it also possesses foliage edged with creamy yellow. Even when the blooms fade, the leaves are good to look at. *P. stolonifera* (south-eastern USA) is a fleshy creeper, more succulent in appearance than its relatives. Its large flowers are held in clusters on stems thrown up from the shoot tips, and the two most popular kinds are pale blue – 'Blue Ridge' – and white – 'Ariane'. They'll grow to no more than 5 or 6in (13 or 15cm) high, but can make a mat 1½ft (45cm) across in a season. *P. douglasii* (western USA) is, they tell us, really *P. austromontana*. Never mind; pick up plants that are varieties of either and you'll find them useful clump-formers of dwarf stature well suited to sink gardens. 'Crackerjack' is a good crimson and 'May Snow' a sparkling white. *P. adsurgens* (Oregon and California) has one curious variety that's worth a seond glance: 'Wagon Wheel' has sizeable, pink blossoms with thin petals, giving them the appearance of the spokes in a cartwheel. Just two more for real spectacle: the two hybrids 'Chattahoochee' – lavender blue with a crimson eye – and 'Daniel's Cushion' – massive deep pink flowers – are vigorous, ground-hugging hybrids that anyone who simply wants to brighten their rock garden should invest in right away. Their late spring and early summer show is dazzling.

Cultivation Generally speaking, these are plants for light soils and bright spots where their blooms can glisten in the sun. But two species – *P. adsurgens* and *P. stolonifera* – do seem to enjoy just a touch of shade and a soil that's laced with peat and leafmould to prevent it from drying out. If you've a suitable patch of earth at the foot of an east or west-facing boulder, try them there. Most species and varieties will tolerate chalk. The varieties of *P. subulata* are, perhaps, the easiest to grow, and they'll thrive on the driest banks if given plenty of water in the early stages to aid establishment. The varieties of *P. douglasii* are fine for sinks and troughs – the others will be too rampant, but look well growing over sunny rock faces and retaining walls. In flower they are breathtaking; out of flower they'll look good as part of an alpine lawn or as a backdrop to winter-flowering bulbs. With age some of the carpets may become a bit threadbare, but regular top-dressing with a sandy, peaty mixture in spring will encourage the production of young shoots and fresh growth to fill the gaps. Replace the old plants only when their performance falls off.

Propagation Cuttings can be rooted in a frame or propagator in early summer. Plants can be divided in spring, and rooted portions planted separately.

Phyllodoce (Ericaceae)

I think most alpine gardeners have a soft spot for *P. caerulea*, for it's a plant of great rarity that's managing to cling to life in one Scottish county. It's distributed fairly widely in other Arctic regions, so there's no need to feel guilty about growing it. Look closely at the plant and you'll see immediately that it's a relative of the heather – the leaves are linear and needle-like and carried in ranks up the erect stems. The flowers are bells of purplish pink, opening in early summer. The whole plant grows no more than 6in (15cm) high. *P.* × *intermedia* 'Fred Stoker' makes a thickly furnished shrublet up to 1ft (30cm) high, generously decorated in late spring with masses of pinkish-purple, urn-shaped flowers. *P. empetriformis* (western USA) has more open bells, each with a clapper-like stigma protruding at the mouth. It grows to around 6in (15cm) high, and its blooms vary in their intensity of pinkness.

Cultivation As you might expect, these are plants for the peat bed – they hate lime in any form and prefer a spot in gentle shade where their roots can delve into a leafy, peaty mixture. Make a pocket for them at the foot of a north-facing boulder if you lack the luxury of a peat garden. Make sure the ground never dries out, and don't plant phyllodoces too deeply – their stems creep through the soil just below the surface. They can also be grown in pots or pans in the alpine house. Pot them up in a mixture of equal parts John Innes ericaceous compost, sieved leafmould and sharp grit, and water them with rainwater, aiming to keep the compost moist all the year round. During most of the year the pots are best plunged in a shaded frame, but at flowering time they can be brought into the alpine house for their season of glory. A daily spray with rainwater will perk up the plants in summer. Repot when necessary after flowering, but top-dress every year with the potting mixture.

Propagation Cuttings of shoot tips can be rooted in a propagator containing equal parts of peat and lime-free sand in summer. Rooted layers can be removed and transplanted in spring, and seeds, when available, can be sown in peat-based seed compost in spring, and germinated in a shady frame.

Phyteuma (Campanulaceae) Rampion

You'll find plenty of rampions with their spiky, burr-headed flowers in various shades of blue poking their way through lush alpine meadows. At home just a handful are offered by nurserymen, among which the pearl is *P. comosum* (now correctly, if less euphoniously, known as *Physoplexis comosa*) (Figure 2.6). It's native to the European Alps, and admired as a stunning curio by nearly all gardeners. From its clump of oval, coarsely toothed leaves there emerges in June or July a cluster of flowers, each one shaped like an elongated minaret, pale lilac at the base and dark purple at the tip where the forked stigma protrudes. On mature plants many of these flower

Figure 2.6
Phyteuma comosa
Devil's Claw

clusters will be carried on 2 or 3-in (5 or 8-cm) stems – in full bloom there's no plant more coveted. My own potful sits in front of me as I write. As yet it's a spry two-year-old with just one stemless flower cluster, but it's still a treat to treasure. After this species all others are a bit of a come-down, but you could try *P. scheuchzeri* (European Alps and Pyrenees). Its royal-blue flowers are carried in clusters atop 12-in (30-cm) stems in late spring.

Cultivation Any ordinary soil – chalky or not – suits most species, and full sun is essential. *P. comosum* loves lime, and if your soil is acid then work in a few limestone chippings to make it feel at home. If you grow it outdoors, find it a sunny crevice between two rocks, or plant it in a sink garden or in a hole bored in a lump of tufa. Wherever it is, make sure that a few slug pellets reside too, for it will soon be polished off by marauding molluscs if you take no precautions. In pots kept in an alpine house or frame the plant grows well in a mixture of equal parts John Innes No. 1 potting compost and coarse grit, plus a sprinkling of ground limestone or limestone chippings. Repot when necessary in spring as soon as you can see the shoots starting into growth. Water well whenever the compost is dry between early spring and autumn, but keep it very much on the dry side from October to February. You'll think you've lost the plant every winter, for it dies away completely above ground. Be patient – if you've kept slugs and snails at bay it will emerge in spring.

Propagation The hardy clump-formers can be increased by division in spring. *P. comosum* is more tricky. Seeds can be sown as soon as they are ripe in late summer. Crush the dried flower and sow the seeds, chaff and all, on the surface of gritty seed compost, covering the seed and chaff with a thin layer of grit. Stand the pot in a frame where it will be frosted during the winter. Keep the compost gently moist and, hopefully, in spring the seedlings will emerge. Prick out individually into pots of the recommended compost as soon as the youngsters are large enough to handle. They will take several years to reach flowering size. The plant is also reputed to grow from cuttings of young shoots taken in spring, and rooted in pots of sandy compost in a frame.

Plantago (Plantaginaceae) Plantain

If you spend half your time trying to eradicate plantains from you're lawn you will quite likely fail to be persuaded to include them in your rock garden. That would be a shame, for there are at least three worth growing for curiosity alone, and they won't really make nuisances of themselves. The trickiest is *P. nivalis*, a Spanish native with rosettes of narrow leaves thickly coated in silvery, silky hairs. Alas, the blooms are just what you'd expect of a plantain, but at least they provide seeds to grow more luscious rosettes. The other two plants I shall mention have leaves suspiciously like the greater plantain, *P. major*. No, don't turn the page, they are really quite handsome and not too invasive. *P. rosularis* is the rose plantain. Its leaves are exactly the same as the related weed, but the flower spikes have become rosette-shaped, leafy bracts of bright green that last for weeks on end. It grows to about 9in (23cm), and is a curio I'd not be without. *P. major* 'Rubrifolia' *is* the weed, but a form with leaves that are a rich burgundy shade of beetroot. It's just a bit taller at around 1ft (30cm).

Cultivation *P. nivalis* really needs frame or alpine-house treatment if it is to survive soggy winters. Its hairy leaves act just like sponges when water hits them, doing nothing at all for the well-being of the rosette. Plant it outdoors in a sunny, sharply drained soil if you must, but have some plants in reserve to make up for losses. It is good in sink gardens, even if only for a season. In pots it delights in a mixture of John Innes No. 1 potting compost and sharp grit in equal parts. Keep it well watered in spring and summer when it asks for a drink, but much drier in winter to avoid rotting. Repot in spring when necessary. The other two plants have robust constitutions and need no coddling at all. Plant them in sunny spots in soils that are reasonably moisture retentive.

Propagation All three come true from seeds, which can be sown in pots of seed compost as soon as ripe, and germinated in a garden frame.

Pleione (Orchidaceae)

Now here's a plant you'll either like or loathe. Alpine enthusiasts are hardly renowned for their love of orchids – those vulgar flowers from tropical jungles – so imagine how they feel when faced with an alpine orchid! Pleiones come from the mountain regions of China and the foothills of the Himalaya, so there's no denying they're legitimate alpines. The one that's most widely available (and probably the only one you'll be able to afford) is that which we must now call *P. bulbocodioides*, from China, Tibet and Taiwan. It embraces a host of species such as *P. limprichtii*, *P. formosana* and *P. pricei* which used to be thought different and distinct. It also has a fair number of varieties with names that indicate their colour or the folk they are named after. All have those classic 'orchid' flowers with a prominent lip or 'labellum' speckled with a contrasting colour, and the ground colour of the flowers may be white or any shade of pink through to deep purplish rose. The charm of this orchid is its size – the blooms sit on stems barely 1in (25mm) long, and open before the leaves unfurl. Spring is its season of glory.

Cultivation These are plants for the alpine house or frame, where they can be overwintered in conditions that are virtually frost free. You can try them outdoors if you wish – in a well-drained, humus-rich soil that is sheltered and protected from winter wet – but I wouldn't give much for your chances of success (or mine, I hasten to add). No; if you want to grow pleiones well, do the job properly. The plants are equipped with onion-shaped 'bulbs', which are correctly called pseudobulbs. In their native environment they are semi-epiphytic, growing in very organic soil on rocks, tree trunks and the like, so they must be provided with a similar medium in cultivation. Don't go to the length of mixing up cow dung and sphagnum moss. Instead, settle for a mixture of equal parts John Innes ericaceous compost, sieved leafmould and sharp grit. This freely draining, yet moisture-retentive mixture can be put into shallow clay pans over a layer of drainage material. Space the pseudobulbs 3in (8cm) apart (if you are lucky enough to have several) or grow one to a 3-in (8-cm) pot. Don't bury the pseudobulbs – just cover their bases and any roots that are present. When potting is completed the surface of the compost should rest 1½in (4cm) below the rim of the container, so that a layer of moss, sieved leafmould or coarse peat can be strewn around the plants to prevent dehydration of the surface and the roots. Potting time is February – before the plants make any growth. Keep the pots or pans plunged in a shaded frame at all times, except when the plants are in flower, then bring them into an alpine house or show frame. As soon as shoots are seen to be growing in March, water can be applied as required – very sparingly at first, then more freely through the summer as the leaves grow. Keep the pans shaded and well ventilated. The leaves will turn yellow and die down during November, and the compost must then be kept very much on the dry side, though not dusty. The pseudobulbs each last only one year, so don't be alarmed when they wither and die, provided that plump new ones are there to take over. Repot only when the existing container becomes overcrowded. Again, February is the best time for the move, though some growers prefer to do the job immediately after flowering.

Propagation New pseudobulbs can be severed from the main clump, and potted up on their own in February or March.

Polygala (Polygalaceae) Milkwort

This is one of those plants that will give you a bit of a start when you see it in flower – its pea-like blooms are a striking combination of magenta and bright yellow. *P. chamaebuxus* 'Grandiflora' (syn. *P. c.* 'Purpurea') is sometimes known as the bastard box, but despite its unpleasant name it's a grand little evergreen shrublet not much more than 4in (10cm) high. The true species is native to Europe.

Cultivation Although it's most at home in a peat bed where gentle shade and a humus-rich medium keep it happy, this is a plant that can be pleased equally by a peaty, leafy soil at the foot of a north-facing boulder. If you've a sink garden in a shady spot try it there – it will be many years before it outgrows its welcome.

Propagation Rooted runners are easily detached and planted up in spring, or cuttings of shoot tips can be rooted in a frame in summer.

Polygonum (Polygonaceae) Knotweed

These are easy carpeters for any rock garden where their rampant habit won't be too much of a nuisance. *P. affine* comes from the Himalaya, and makes a thick mat of oval leaves that turn rich coppery brown in autumn, remaining on the plant right through the winter until they are masked by fresh spring growth. The tiny, pink flowers are carried in spikes at the top of 9-in (23-cm) stems right through summer. It is nearly always recommended that the two varieties 'Darjeeling Red' and 'Donald Lowndes' be grown in preference to the true species. They are both superior in flower, having rather darker blooms, but my experience of 'Darjeeling Red' is that it does not colour up its foliage so well as the true species in autumn. Pay your money and take your choice – one of each perhaps? There's one more species you should grow if you've room. *P. vacciniifolium* is a smaller-leaved, more slender-spiked version of *P. affine*. It hails from the same terrain, but doesn't flower until late summer. When it does it is obliging enough to continue well into autumn, turning its leaves brown afterwards. Through the growing season some of its leaves will show their grey undersides to make a pleasant contrast to the darker green of the upper surface.

Cultivation I grow both just above bare-faced boulders on a rock bank so that they can lay their carpet of stems down the sheer face to show off both flowers and foliage. As to soil they are unfussy, putting up with chalk and anything else that's well drained and reasonable. They worship sun. Try them on top of retaining walls – or even in the sides – and alongside steps. They also thrive in gaps between paving slabs on a patio. They more than earn their keep.

Propagation This is dead easy – by division in spring.

Potentilla (Rosaceae) Cinquefoil

Dozens and dozens of potentillas adorn nurserymen's lists, and it's rotten of me to hack my selection down to three. But they're three good ones. The commonest is undoubtedly *P. fruticosa* (Northern Hemisphere), which has numerous varieties. It's certainly rather too large for miniature rock gardens, but where there's a yard (metre) to spare, then it will pay for its keep by flowering off and on from May to September. You'll have to be guided by what's available from your own pet nurseryman, but the following varieties will not disappoint you: 'Elizabeth', primrose yellow; 'Tangerine', orange; 'Katherine Dykes', pale yellow; 'Tilford Cream', white. Most of the plants grow to just under 3ft (90cm) high, but 'Tangerine' and 'Tilford Cream' will usually grow to between 1 and 1½ft (30 and 45cm). All have arching stems and those lovely, deeply cut leaves that look like green stars. You'll have to be a gambler to try the two red varieties 'Red Ace' and 'Royal Flush'. In the right situation (light shade) and with the right weather conditions – cool, dry weather – the flowers are very showy. But the plants take time to settle in and also vary in their performance on different soils, producing wishy-washy blooms if all is not well. It's always struck me as rather a hit and miss affair, and I'd sooner plump for a white, red or orange variety *knowing* that it's going to do well. If you are prepared to take a chance, then take one with *P. nitida* 'Rubra' from the Dolomites. It makes mats of silvered cinquefoil leaves that are topped with rich rose-pink flowers in early summer. The flowers are virtually stemless, making this a compact plant of great attraction. *P.* × 'Tonguei' is my final choice. It spreads flat on the ground from a central crown and at the end of its stems in the heart of summer erupt apricot flowers centred with crimson. The leaves are rich green, often burnished bronze.

Cultivation Most potentillas are remarkably accommodating, growing well in any reasonable soil – chalky or not – where drainage is anything except dreadful. They appreciate sun, but will tolerate light shade. Try the shrubby types in sunny beds or banks facing any direction. *P.* × 'Tonguei' wants to be positioned where you can look down on it easily. *P. nitida* 'Rubra' can best be persuaded to grow in a scree bed where the poor diet, brisk drainage and good light will be much appreciated. It will also thrive in sharply drained sink gardens or holes bored in tufa. If all you have to offer is a rock garden, find it a sunny crevice. In the alpine house it will do well in pots containing a mixture of John Innes No. 1 potting compost and sharp grit in equal parts. Overfeed and overpot it and it will be shy of flowering. Water freely when necessary when the plant is growing; keep it on the dry side in winter – but not dusty. Repot only when absolutely necessary in spring.

Propagation Cuttings of shrubby types can be rooted in a propagator in summer. Clump-formers can be carefully divided in spring, and either potted up (in the case of *P. nitida* 'Rubra') or planted out (in the case of *P.* × 'Tonguei').

Pratia (Campanulaceae)

These are a couple of carpeting curiosities. *P. angulata* 'Treadwellii' (New Zealand) makes a thick mat of tangled stems with indented, fleshy green leaves and is decorated in summer with white, lobelia flowers. Nothing curious yet; it's when the red fruits form that the plant becomes more remarkable. Shaped like tiny onions, they are burgundy red and stud the foliage for many weeks. *P. pedunculata* (Australia) is mentioned hardly anywhere in alpine literature, but it has recently crept into cultivation, and in spite of its origins is proving to be quite hardy. It makes fresh-green, ground-hugging carpets much neater than the previous species, seldom reaching upwards more than 1in (25mm). The starry flowers are pale blue, and nestle among the foliage in spring and summer. When it flowers well it's a real treasure.

Cultivation Sun or light shade and a soil that's not likely to dry out in summer are the plants' requirements. I lost *P. angulata* 'Treadwellii' on sandy soil through carelessness and infrequent watering in the early stages, but it's not a difficult plant given the right spot. *P. pedunculata* now thrives on a north-facing slope in chalky soil, as would its relation. Try them also in gaps between paving stones and in alpine lawns. Keep them out of sink gardens for they'll swamp your tiny treasures.

Propagation Mature plants can be divided in spring, and the divisions planted out at 9-in (23-cm) spacings if a thick carpet is wanted. Cuttings can be rooted in a propagator in summer. Seeds can be sown in pots of seed compost in spring, and germinated in a frame or cool greenhouse.

Primula (Primulaceae) Primrose

The primulas are a devastatingly beautiful tribe composed of hundreds of species from Asia, Europe and America. Some are tall and stately, others are dainty miniatures. I'm a confirmed fan of these plants, but cannot ramble on as long as I'd like to about their different qualities – instead I'll mention the most widely grown types, giving details of their own peculiar cultivation requirements.

Cultivation In the garden the primulas that makes the most impact are undoubtedly the candelabra types, with their 1½ to 2-ft (45 to 60-cm) stems of tiered flowers in shades of red, orange, yellow and pink. For the gardener who wants a good mixed bunch, those grown from seed supplied by the Northern Horticultural Society and labelled 'Harlow Car Hybrids' take some beating (see plate section). Of the species, try *P. japonica* (Japan), deep purple-red; *P. helodoxa* (Burma, Yunnan), yellow; *P. bulleyana* (Yunnan), orange-red, and *P. pulverulenta* (Szechuan), rich pink. All this little lot love a sunny or lightly shaded spot in a soil that has been amply laced with peat, leafmould or compost so that it never dries out. They are happy, too, at streamsides, showing their preference for a moisture-retentive soil. When suited they'll seed themselves around freely. Some gardeners find them difficult on chalk; others have succeeded with them – much depends on the site and the way the soil is enriched. *P. sikkimensis* (Himalaya), with its drooping bells of pale yellow carried on stems as high as the candelabra types, is a bashful beauty that appreciates the same conditions. *P. vialii* (Yunnan, Szechuan) captivates all who see it. From a tuft of downy primrose leaves a pointed flower spike emerges. The blooms are crimson red in bud and lilac when open, and as they open from the bottom upwards the effect is of a miniature red hot poker spike, though much more elegant and quite differently coloured! Grow it like the candelabras (though not in really sticky soil), but be prepared for it to have a short life. All these plants will appreciate a spring mulch of organic matter – peat, pulverised bark or leafmould – and they bloom in late spring and early summer.

Our common primrose, *P. vulgaris*, is blissfully happy in most moisture-retentive but not soggy soils – even chalky ones – in sun or dappled shade. Don't gather it from the wild; several seedsmen now sell the seeds, which are not at all difficult to grow. The hybrid 'Garryarde' types are becoming tremendously popular again, having fallen out of favour years ago. Their leaves are flushed with mauve, and their blooms are dusky-shaded primroses of lilac, pink or pale mauve. Almost as common as the primrose is *P. denticulata* (China, Himalaya, Afghanistan), the drumstick primula. Its spherical flowerheads of rich mauve, pale lilac or white are carried on 9-in (23-cm) stems in spring. Again, any moisture-retentive soil suits it, and a spot in sun or dappled shade.

The auriculas, *P. auricula* (European Alps) are traditional cottage-garden flowers very much at home on the rock garden in a shady pocket of rich, leafy soil. Choose the types with flowers that are coated in white meal or green farina if you want conversation pieces. The ground colour of the blooms may be red, yellow or white. Grow the plants in pots to brighten an alpine-house display with spring flowers. The plants are happiest in a mixture of equal parts John Innes No. 3 potting compost and sharp grit. Repot or top-dress each spring, and water freely when necessary from spring to autumn, more carefully in winter, but don't let the compost dry out.

If you've a patch of earth that's almost mud, there's one primula which will relish such damp conditions and that's *P. rosea* (Himalaya and Afghanistan). In very early spring it will send its flower shoots through the soil to open their rich-pink blooms well before the leaves unfurl. Eventually it will make 1-ft (30-cm) stems, but not for several weeks. *P. clarkei* (Kashmir) is a miniature relative of *P. rosea* with even more delicately cut blooms of a gentler shade of pink. It, too, likes moist soil (though not, perhaps, as muddy as its relation), and the secret of succeeding with it is to divide it every couple of years to keep up its strength.

An easier dwarf is *P. juliae* (Caucasus), a tiny primrose with magenta-purple blooms that nestle among the rich-green leaf rosette. A good soil in gentle shade will be preferred to dusty earth in full sun.

Finally to the enthusiast's primulas: *P. allionii* (Maritime Alps) is a treasure that's best grown in pots in the alpine house or frame. It studs its tight rosettes of tacky, felted leaves with wide-faced flowers of pink, mauve or white in early spring, and is the delight, or despair, of growers. There are a number of varieties, but I think the one sold as 'Crowsley Variety' is one of the best and most robust. Its blooms are a rich shade of magenta. Grow the plants in a compost consisting of equal parts John Innes No. 1 potting compost and sharp grit, with an extra dusting of ground chalk or a few limestone chippings added. Don't overpot – the plants increase very slowly. A good top-dressing of chippings is essential to prevent the collar of the plant from rotting, and water should be applied carefully at all times, though freely when necessary from spring to autumn. I water mine from above and they've not yet come to any harm. Keep the compost very much on the dry side in winter, but not dusty. Remove any faded flowers immediately they are seen, and pull out dead leaves with a pair of tweezers, once in June and again in December. Repot after flowering every other year. Failure sometimes occurs as a result of meanness with the watering can in summer; make sure the compost is gently moist at that time, and shade the plant from direct sunshine by plunging it in a shady frame.

The petiolarid primulas are at their best in early spring, and are drooled over by peat-garden enthusiasts. They are typified by such plants as *P. gracilipes* (Bhutan, Tibet, Sikkim, Nepal) with its thickly packed rosette of crispy green leaves and short-stemmed, pinkish-blue flowers. *P. whitei* (syn. *P. bhutanica*) (Himalaya, Assam, Bhutan, Tibet) has blue flowers centred with white, and there are others of similar shades, plus jewels like *P. aureata* (Nepal) with yellow, stemless flowers. It's rare, expensive and much sought after, as well as tricky to grow. To generalise about these plants, they all relish peat-bed conditions – a shady spot in a peaty, leafy soil that's not likely to dry out at any time, and they all do best in the north of England and Scotland, where the cool, moist climate is more to their liking. Guard against slugs at all times. Grown in pots, the petiolarids are happy in a mixture of equal parts John Innes ericaceous compost, sieved leafmould and sharp grit. They are best kept plunged in a north-facing frame until they bloom, when they can be moved into the alpine house or show frame for their season of spectacle. Take care, though, to keep them out of bright sunlight. Water them freely in spring and summer to keep the compost evenly moist at all times. Keep it just damp – not soggy – in winter. Remove the dead foliage in winter. Protect from severe frosts by covering the frame with lights and sacking or carpet in winter. Repot annually after flowering.

Propagation Generally speaking, any primulas that form clumps can be divided. Do this job in autumn for the late-spring and early-summer bloomers such as the candelabra types. The very early bloomers, such as the petiolarids, can be divided as the flowers fade. *P. denticulata* can be propagated by division or seed, but also by root cuttings inserted vertically in sandy soil in a frame in spring. Many primulas are easiest and cheapest to grow from seeds, which can be sown either as soon as they are ripe (which is preferable), or in early spring, in pots of peat-based compost. Germinate them in a garden frame. *P. allionii* is best propagated by tiny cuttings of individual rosettes dibbed into pots of pure, sharp sand in spring. Root such cuttings in a cool, shaded frame.

Pulsatilla (Ranunculaceae) Pasque Flower

Far from shunning their native plants, most alpine growers have a soft spot for them. So it is in Britain with the pasque flower, *P.*

vulgaris, sadly becoming rarer by the day in its southern and eastern England haunts. But plants are still widely available from nurserymen, and on the rock garden or bed they'll brighten early spring with their nodding, full-petalled blooms swathed in a coat of silky hairs. The petals are rich purple on the inside, and centred with a knob of bright-yellow stamens, though there are also forms with ruby-red petals and some that are glistening white. Eventually the stems will reach 9in or 1ft (23 or 30cm), and as the flowers fade they turn into fluffy spheres of seeds. The carroty leaves push up after the flowers and support the seedheads on a green fuzz of growth. *P. alpina* is another European alpine, this time with pure-white blooms, downy on the outside of the petals and flushed with pinkish mauve. The downy leaves are more attractive than those of *P. vulgaris*. And finally, *P. vernalis* (Europe), perhaps the finest of the lot, which is not always freely available, but when it is, make sure of obtaining it. The wide, white, cup-shaped flowers sit on 1 or 2-in (25mm or 5cm) stems over the roughly cut foliage, which persists to some degree in winter. Each petal is rich mauve on the outside and clad in the usual silky hairs; a little treasure. All the pasque flowers bloom in mid to late spring, and all were previously included in the genus *Anemone*.

Cultivation Good plants for idle gardeners, the pulsatillas resent disturbance once they are established, and so can be left alone. Plant them in open, sunny spots where the soil is well drained and preferably enriched with leafmould and a few limestone chippings – the plants adore chalky soil (except a pale-yellow-flowered version of *P. alpina* known as 'sulphurea', which seems to resent alkaline conditions. Some gardeners have also found difficulty on chalk with *P. vernalis*, but I'd take a chance if I were you). The younger the plants are at planting time, the more likely they are to establish without trouble. Pot-grown plants are essential.

Propagation Seeds can be sown in pots of seed compost as soon as ripe, and germinated in a garden frame. Transplant the seedlings to individual pots as soon as they can be handled. They will take several years to reach flowering size. Especially desirable forms of *P. vulgaris* can be increased from root cuttings, though the parent plant will resent the disturbance necessary to carry out the operation. Sever 2-in (5-cm) long sections of the thickest roots in July or August, and pot them vertically (the right way up) in pans of sandy compost placed in a frame. The plants that arise from the roots should be ready for potting up in late summer or early autumn.

R

Ramonda (Gesneriaceae) Rosette Mullein

Now here's a beauty. There are several species of *Ramonda*, but only one is offered with any frequency and that's *R. myconi* (syn. *R. pyrenaica*) from the Pyrenees. It makes a flat rosette of wrinkled leaves coated in coarse hairs, and from the centre of each mature rosette a stem or two of large, pale-purple potato flowers will emerge in summer. Yellow anthers stud the centre of each bloom. The flower description should not make the plant sound mundane, for it's a superb alpine that every enthusiast should grow.

Figure 2.7
Ramonda myconi
Rosette Mullein

Cultivation Essentially a plant for north-facing crevices between rocks or in retaining walls of banks or raised beds, ramonda likes a soil that's well drained yet not lacking in humus. Work in both gravel and leafmould or peat for best results. It's not an easy plant to insert after you've built a wall or rock outcrop, so try to plant it as building progresses. The rosette should be positioned at right angles to the ground so that it sheds all surplus moisture. It might look to you as though the plant is uncomfortable on its side in a shady spot, but that's what is likes best. Small plants may take a year or two to flower. The plant may also be grown in pots plunged in a shady frame, and brought into the alpine house when in flower.

Pot it in a mixture of equal parts John Innes No. 1 potting compost and sharp grit, and wedge a couple of stones under the crown to keep the leaves clear of the compost and the surface chippings. Water freely when necessary during the growing season, more sparingly in winter, keeping the compost just moist. Remove any rotting leaves with a pair of tweezers. Repot after flowering when the container has been outgrown.

Propagation Mature plants can be divided carefully in spring. Seeds can be sown on the surface of a peat-based compost in spring, and germinated in a cool propagator. Leaf cuttings with as much stalk as possible can be rooted in a mixture of peat and sand in a cold frame in summer.

Ranunculus (Ranunculaceae) Buttercup

You'll often hear the lesser celandine, *R. ficaria*, described as a pernicious weed, which indeed it is when it is let loose among choice plants of weak constitution. Plant it where it can weave among larger plants, though, and you'll not find its habit quite as annoying. It grows, not through my introduction, in a border in my garden, and each spring it is a sheer delight. Then, just as I'm beginning to wonder whether I should really yank it out, it dies down, leaving the soil quite bare – invisible until the following spring. This does mean that summer bedding plants and the like can then be planted where the celandine sleeps (for it's dormant by June), but on the rock garden the system won't work quite as well. The answer is to plant the choice forms of *R. ficaria* such as 'Albus', white; 'Aurantiaca', orange, and 'Flore Plena', with double blooms. They're not so invasive, and so do not leave large and embarrassing gaps when they die down. *R. parnassifolius* (European Alps) is a challenge. In its best forms it makes a plant 6in (15cm) high with full-petalled, white flowers held in clusters at the stem tips in late spring. The leaves are dark green, glossy hearts forming a tuft from which the flower stem rises. Poor forms will have thin petals or no petals at all, so make sure that you buy a proven beauty. *R. calandrinioides* (Atlas Mountains of North Africa) also varies in its flower quality. The white, goblet-shaped blooms are tissue thin, often flushed pink, and held on 6-in (15-cm) stems over oval, grey-green leaves at any time from early winter to early spring. Buy your plant from a nurseryman known to carry good stock to make sure that it has a full-petalled bloom, not one with narrow petals that give it a scrappy appearance. Many alpine enthusiasts are now growing *R. asiaticus* (south-west Asia and Crete) in pots for an early spring display. Their round, papery blooms of white, pink, red, mauve or yellow are carried on 1 to 1½-ft (30 to 45-cm) stems over typical buttercup leaves. Cultivation in pots is as for *R. calandrinioides*, but keep an eye open for greenfly.

Cultivation The forms of *R. ficaria* are happy in any reasonably moisture-retentive soil in a sunny or shady pocket on the rock garden or at the front of a flower bed. *R. parnassifolius* is a lime-lover, and is happiest on a limestone scree where its roots can delve into the hospitable mix. It relishes full sun. Alpine-house enthusiasts can grow it in pots of a compost consisting of equal parts John Innes No. 1 potting compost and sharp grit, plus additional limestone chippings or ground chalk. Water freely in spring and summer when dry, but keep very much on the dry side in winter until growth is evident. Repot after flowering when the existing container is outgrown. *R. calandrinioides* flowers so early that its blooms are often damaged by wind and rain before you set eyes on them. Grow it in a sunny, sheltered spot in good but well-drained soil on the rock garden, rock bed or scree if you like, or make sure of success by cultivating it in pots for an alpine-house or frame display. A compost of equal parts sharp grit and John Innes No. 1 potting compost will suit it, and it is a plant which needs a resting spell. In early summer the leaves will die down, and the pot should be placed in a sunny frame and completely deprived of water until early autumn. Then the compost should be knocked out, the crown salvaged and repotted in fresh compost so that it sits just below the surface. Water very carefully now until the shoots emerge (take care not to saturate the compost, simply keep it evenly moist). Protect the plant from frosts at this stage. Water freely when necessary as the season advances, and keep the plant well ventilated whenever possible. When the blooms appear the plant can be placed in the alpine house, and as its season ends it can be returned to the frame prior to being dried off.

Propagation *R. ficaria* is easy to propagate by division in early spring. *R. calandrinioides* can be divided carefully at repotting time, or seed sown in pots of seed compost in spring, and germinated in a frame. *R. parnassifolius* is also propagated from spring-sown seeds.

Raoulia (Compositae)

The gardener is conditioned to dislike mosses, which is a pity, for they are handsome plants in the right spot. Never mind, he can grow raoulias instead. These are the most ground-hugging carpeters you'll ever see, pressing their minute rosettes of evergreen foliage close together and close to the soil to form a green or silver sheet just like a gigantic pad of moss. The greenest species are *R. glabra*, *R. lutescens* (syn. *R. subsericea*) and *R. tenuicaulis*. The first-named produces tiny, whitish flowers, the others yellow ones in summer, but these hardly mar the spectacle of the foliage. For a silver blanket try *R. australis*, which is not quite so reliably hardy as the others, or *R. hookeri*, with slightly larger silver rosettes. These two need a bit more cherishing, but will amply repay you for any trouble you take by spreading well. They all come from New Zealand.

Cultivation Full sun and brisk drainage is what raoulias need. Grow them in sink gardens, on scree beds, in lumps of tufa over which they can grope their way, and in crevices and gritty pockets on the rock garden. No plant is better equipped to make an alpine lawn – let other plants shoot through it so that it can act as a verdant or silver backdrop to their contrasting foliage and flowers. Take a little care over watering the plants, even in the open. I've lost them on very dry soil in summer, but could have saved them had I thought to give them a drenching now and again. In the alpine house or frame the plants can be grown in shallow, wide-rimmed pans of a compost consisting of equal parts John Innes No. 1 potting compost and sharp grit. Give them good light at all times, and water them freely when necessary in summer, more sparingly in winter so that the compost remains much drier but not dusty. Repot in spring when necessary.

Propagation Divide mature clumps in late summer.

Rhodohypoxis **(Hypoxidaceae)**

In every gardener's collection there are 'ooh!' and 'aah!' plants – the ones that always draw admiring comments from spectators. This is one of them. *R. baurii* comes from the Drakensberg Mountains of South Africa, and though it is hardy in well-drained rock gardens, it is most often seen as a pot plant in the alpine house. Its leaves are long, narrow and hairy, growing to a height of around 3in (8cm). Among them, at any time from spring to late summer, are carried plenty of broad-petalled, starry flowers of light or dark pink, white or almost crimson. There are lots of named varieties to choose from, few of which will disappoint you. All deserve their 'oohs' and 'aahs'.

Cultivation If you don't possess an alpine house or frame, do give the plants a try outdoors. Find them a sunny, sheltered spot where the soil can be enriched with a helping of peat and plenty of sharp sand or grit. This well-drained but moisture-retentive medium is what they like best. They don't like chalky soil. In summer they need to be prevented from baking, and in winter guard against waterlogging – not an easy task. In pots the plants will relish a mixture of equal parts John Innes ericaceous compost and lime-free grit. Water freely when necessary during spring and summer, but when the foliage dies down in autumn withhold water completely until the following spring. Repot annually just as growth commences.

Propagation Divide clumps during the growing season.

Roscoea (Zingiberaceae)

These weirdos are intriguing plants to grow, sending up succulent sheaths of green leaves from which emerge bearded flowers of yellow, pink or lilac-purple. All come from China and the Himalaya, and the most freely available species are *R. cautleoides*, 9 to 12in (23 to 30cm), pale yellow, and *R. purpurea*, 6in (15cm), lilac-purple in summer. If you're ever offered *R. alpina*, give it a try. Its flowers are of a more feeble constitution than the other species, but at 4in (10cm) high the plant has a certain daintiness. The blooms are pink.

Cultivation Traditionally recommended for semi-shady spots, roscoeas will be absolutely fine in full sun if they have plenty of moisture at the roots. Work in sand and grit to improve drainage, but also lashings of peat and leafmould to retain moisture in summer. I've not tried them in my chalky soil, but some gardeners have succeeded on chalk where the ground has been enriched with peat and leafmould. Prepare the earth well and you can sneak roscoeas in almost anywhere, except in really heavy shade.

Propagation The fleshy roots can be divided and replanted 4in (10cm) deep in spring. Some plants will seed themselves successfully. Seeds can also be sown in peat-based seed compost in a frame in spring. The seedlings will take about three years to reach flowering size.

Sagina (Caryophyllaceae)

There are not many saginas, but I certainly wouldn't want to do without the two species I grow. *S. boydii* (originally from Scotland) makes a tight, tiny dome of the darkest, glossiest, green leaves you'll ever see – they look as if they've been varnished. It seldom grows more than an inch or two (25mm to 5cm) high, and increases laterally very slowly. *S. glabra* (syn. *Arenaria caespitosa*) (northern

Europe) is not so circumspect, but is equally desirable in its golden, filigree form 'Aurea'. This plant makes a tufted mat of fine, acid-green candyfloss, studded in summer with tiny, white stars. It's another ground-hugger, and will feel its way over and around flat boulders and gravel.

Cultivation Both plants love sun, and *S. glabra* 'Aurea' doesn't seem to mind chalky soil. Grow it at the foot of a rock bank or the front of a bed or as part of an alpine lawn. Find *S. boydii* a spot in moisture-retentive soil in a sink or trough or a lime-free scree bed. It is reputed to dislike chalk, but grows happily for me in a mixture of equal parts John Innes No. 1 potting compost and sharp sand. Try it in pots of the same mixture in an alpine house or frame. Water it when necessary in summer; keep it slightly drier in winter. It is adored by slugs (I speak from bitter experience) and also, apparently, by red spider mites (I keep my fingers crossed). Don't worry about any apparent foliage burning in winter; it will soon recover. Repot in spring only when the existing small pot is outgrown.

Propagation *S. glabra* 'Aurea' can be divided in spring. Cuttings of *S. boydii* consisting of single tufts can be rooted in pots of sand in a propagator in early summer.

Salvia (Labiatae) Sage

Every now and again you need a really bold plant to offset some of the fussy types that are common in the alpine world. One such plant is *S. argentea* from eastern Europe. It has the physical grandeur of *Bergenia* (elephant's ears) made more acceptable by its thick coating of white hairs. These glisten with dewdrops in spring and early summer, and only lose their dense wadding when the flower spike emerges later in the year. The spike is best pulled off plants grown on the rock garden, for it pushes upwards to a couple of feet (50cm). You can also pull off any coarse, green leaves to let the downy youngsters see the light. In spite of its late-season abandonment, it's still worth growing for those early leaves. *S. caespitosa* (Turkey) is altogether smaller, but equally handsome in a more restrained fashion. It makes short, woody stems just 2 or 3in (5 or 8cm) high, clad in aromatic, downy, grey leaves that are finely cut. It's decorated in summer with inflated pink flowers, conspicuously hooded as you'd expect in this family, and is a little beauty.

Cultivation Brilliant sun and any well-drained soil on a rock garden, rock or scree bed is what these plants need, though *S. caespitosa* is really best grown in pots in an alpine house or frame, where it can be protected from severe cold and wet in winter. In spite of this I lost my plant in the hard winter of 1981/2. It enjoys a compost consisting of equal parts John Innes No. 1 potting compost and sharp grit. Water it carefully at all times, keeping water off the foliage, and keep it almost bone dry in winter (but not quite!). Repot in spring when necessary. Snip out any dead growths as soon as they are seen. *S. argentea* is usually biennial and almost always short-lived, even if its flower spikes are removed as soon as they are seen.

Propagation Both species can be propagated by seeds sown in spring in pots of seed compost, and germinated in a frame. *S. caespitosa* can be propagated by heel cuttings inserted in sandy compost in pots in summer. Place the pots in a shaded frame.

Sanguinaria (Papaveraceae) Bloodroot

The double-flowered bloodroot, *S. canadensis* 'Flore Pleno' is a plant to make you drool. From the bare spring earth emerge miniature 'chrysanthemum' flowers of purest white, set off well against the scalloped leaves of purplish grey, then greyish green. Sadly, the flowers are fleeting, dropping their petals in a couple of days; but what a spectacle while they last! The true species is single flowered and even more fleeting, and as its name suggests it comes from Canada. The double one beats it every time, I reckon.

Cultivation This is a plant that relishes a peaty, leafy soil that never dries out. Find it a home in a peat bed or a deep, leafy soil in a rock garden pocket where it will also have the shade from bright sun that it insists on. Try to avoid disturbing it once it's planted.

Propagation Divide mature plants as carefully as possible in late summer when the foliage dies down.

Saponaria (Caryophyllaceae) Soapwort

The rock soapwort, *S. ocymoides*, is one of those sterling alpines to bracket with aubrieta, alyssum and cerastium, though it is rather less rampant than all three. It comes from the European Alps, and both there and in gardens it smothers itself in round, pink flowers during summer, tumbling in orderly clouds from sheer rock faces and banks. Nurserymen offer a white form 'Alba' and a neater, darker form which is sold as 'Rubra Compacta'. None of them is very long lived (flowering themselves to death I suspect), so always grow a few youngsters to replace the pensioners. *S. caespitosa* (Pyrenees) makes more of a tufted plant with rich-pink flower clusters held on 4-in (10-cm) stalks. The leaves are quite narrow. A hybrid between these two species, *S.* × *boissieri* forms a tidy cushion with pink flowers on 3-in (8-cm) stalks. All three are jolly little plants, well worth squeezing in for a bit of rosy colour.

Cultivation Sun worshippers all, the soapworts should always be planted in full light. Place *S. ocymoides* so that it can tumble down a bank, over a sheer rock face or over a retaining wall. It likes any well-drained soil, and doesn't care whether it's chalky or not. *S. caespitosa* and *S.* × *boissieri* are most at home in sunny scree beds or sink gardens, where they'll not prove too rampant.

Propagation Cuttings of non-flowering shoots can be inserted in sandy compost in a frame in summer, and the young plants potted up and grown on prior to being planted out. *S. ocymoides* produces seed which can be sown in spring in pots of seed compost, and germinated in a frame.

Saxifraga (Saxifragaceae) Saxifrage, Rockfoil

In moments of despair at my ignorance of the enormous number of garden plants, I often think I'll specialise in saxifrages and become an 'expert'. But a glance at a comprehensive reference work on the plants quickly dispels any hope I might have of reaching erudite peaks with rapidity – there are thousands of them. If this is a disadvantage when trying to become an expert on saxifrages, it's certainly an advantage when growing them. I've seen vast trough gardens planted up solely with saxifrages, and still offering variety and interest through the year; for as well as being breathtaking in flower, they are neat and eyecatching in leaf too. Any normal gardener is bound to be put off the plants by the vast and unwieldy system under which they are classified. They are divided into fifteen different groups. Forget them. For most folk's purposes it's sufficient to remember just four different types: mossy saxifrages; Kabschia or cushion saxifrages; encrusted saxifrages, and 'others'.

First the mossies. These are the commonest type and the easiest to grow, judging by their ability to survive in neglected rock gardens and stone heaps countrywide. They make low, spreading mounds of fresh green rosettes, springy to the touch and often burnished with red or plum purple – especially in cold weather. The flowers are starry, carried in clusters on 3 or 4-in (8 or 10-cm)-high stalks in spring, and they may be white, yellow, pink or rich rosy crimson. Eventually the mats will grow 2ft (60cm) or more across, sheeting themselves with bloom. Nearly all mossy saxifrages are hybrids with descriptive or commemorative names. Choose them in flower to be sure of finding the best.

The Kabschia saxifrages flower early in spring, before the mossies. They are little treasures. The leaf rosettes are very hard and tightly packed together, so that when the cushion is pressed it hardly gives at all. Depending on the species the rosettes may be fresh green, grey green, blue grey, and decorated with a little white meal (though not usually as generously as the encrusted saxifrages). Here the flowers are carried on stalks 2 or 3in (5 or 8cm) high, often singly, but sometimes in small clusters. They may be white, yellow, pink or mauve, and often have dark red stalks. The mounds of foliage increase quite slowly – nothing like as fast as the mossies. Here I will give you some species and varieties – just the ones I've found among the best: *S.* × *apiculata*,* pale yellow; *S. burserana* 'Gloria', large, white flowers on red stalks; *S. grisebachii* 'Wisley Variety', large rosettes of grey-green leaves from which rise shepherd's-crook stems densely coated in bright-crimson hairs; *S.* 'Irvingii', pink; *S.* 'Jenkinsae',* pink; *S.* 'Boston Spa', primrose yellow; *S.* 'Valerie Finnis', large, primrose-yellow blooms; *S.* 'Elizabethae',* clear yellow; *S.* 'Winifred', rich mauve. (* – especially suited to outdoor cultivation).

The encrusted saxifrages might not at first glance possess the flamboyance of the two previous groups, but they are certainly more elegant. Their leaves are usually elongated, making the rosettes larger and more starry, and the wiry, branching flower stems often arch gracefully. The flowers are starry too – white, pink or white spotted with crimson – and carried in airy clusters in spring or summer. These are the saxifrages offering most year-round interest, for their foliage is always good to look at, being encrusted round the edges with a white, limy deposit. They don't usually rise up from the ground more than an inch or two (25mm to 5cm), but many of them can spread over a foot (30cm) or more. Of the ones I grow I can recommend: *S. cochlearis* (Maritime Alps) in its form 'Minor', with small, tight-packed rosettes and white flowers on 5-in (13-cm) stems; *S. cotyledon* (Northern Hemisphere) with broad-leaved rosettes and dense flowerheads of white up to 1ft (30cm) tall. *S. longifolia* (Pyrenees) and *S.* 'Tumbling Waters' are real gems with single rosettes as handsome as sea anemones. When the rosette is mature a vast, much-branched flower stem arches from the centre, to display a rich array of white flowers. The true species dies after flowering (hopefully setting seed), but the hybrid will produce offsets. *S. paniculata* (syn. *S. aizoon*) (Europe) I grow in its forms 'Lutea', with pale-yellow flowers; 'Rosea', with pale-pink blooms and 'Balcana' with white flowers spotted with red. The encrusted rosettes of all are small and dense packed, and the flower stems are around 6 to 9in high (15 to 23cm). *S.* 'Southside Seedling' is a must for your rock garden. Its rosettes are as large as those of a houseleek, fresh green and edged with rime. From the centre of each comes a stem over 1ft (30cm) long, from which explode the starry, white flowers heavily blotched with crimson. It's a grand sight.

Figure 2.8 *Saxifraga oppositifolia* Purple Saxifrage

And so to 'the rest'. I've just three to recommend here, but what a trio. The first is *S. oppositifolia* (fig. 2.8) – the last saxifrage I'd part with from my collection, probably because it's found wild on Pen-y-ghent in my native Yorkshire. But, sentiment aside, it's a treat from early to late spring, when its strands of succulent, tiny, green leaves are studded with stemless mauve-purple flowers. The plant forms a loose, spreading mat, and has several good forms of which the most freely available is 'Splendens'. Try also 'Latina' and 'Ruth Draper' – the latter is said to have slightly larger flowers of paler pink, but I've seen considerable variation in both colour and size of plants labelled 'Ruth Draper' on the showbench. Don't bother with *S. oppositifolia* 'Alba' – it really is a feeble-flowered weakling compared with its deeper-coloured relative. The true species, as well as coming from Yorkshire, is also found in other parts of northern Britain and the Northern Hemisphere.

London pride, *S.* X *urbium* (syn. *S. umbrosa* of gardens) should not be excluded from the rock garden just because it is common. If you really find its spoon-leaved rosettes and airy plumes of flowers too large for your situation, go for one of its smaller varieties such as 'Elliott's Variety'. *S.* X *urbium* 'Variegata Aurea' has leaves marbled with butter yellow (though it might look too virus-like for some).

The latest-flowering saxifrage in my little bunch is *S. fortunei*

(China, Japan, Korea, Manchuria). It makes clumps of deeply lobed, glossy leaves of fleshy texture that are held on bright red, hairy stalks. The foliage is burnished with bronze as it matures, but for spectacular year-round effect choose the variety 'Rubrifolia'. The large, much-branched flowerheads open in autumn, revealing neat, white blooms with elongated lower petals – comets instead of stars. The whole plant will grow to around 1 or 1½ft (30 or 45cm).

If you want to delve deeply into the world of saxifrages, get hold of a copy of Winton Harding's *Saxifrages*, published by the Alpine Garden Society. It's a readable masterpiece guaranteed to satisfy your curiosity.

Cultivation As a race the saxifrages are not difficult to grow, but failures have been known – usually when the plants have been positioned in sun-baked spots. It may seem rather surprising that nearly all saxifrages prefer dappled or very gentle shade to full sun, but it's true nevertheless. For ease of reference I've split up my cultivation hints so that they are applicable to the different groups.

Mossy saxifrages are certainly happier in stiffer soils, (chalky or not), rather than those of a light, sandy nature, where they tend to develop brown patches in summer. In moisture-retentive earth they will tolerate full sun, but it's safer to plant them in light shade to be on the safe side. Not only will this prevent browning, but it will also stop the red-flowered varieties from fading fast. Mossies spread rapidly, so site them where they can be allowed to roam freely. They are good front-of-the-bed plants, are exquisite in alpine lawns and enjoy tumbling among rocks. Divide them at the first sign of scrappiness and prolong their lives by top-dressing them with John Innes compost each year after flowering.

Kabschia saxifrages are nearly always seen at their best in the alpine house, where they can be grown in pots containing a mixture of equal parts John Innes No. 1 potting compost and sharp grit, plus a few limestone chippings. Blooming very early in spring (or even late winter) their blooms and cushions can suffer in foul weather outdoors, but I still recommend that you grow a few Kabschia hybrids outdoors to start the season early. Those which are asterisked in the descriptive section I have seen growing well on rock beds and gardens and in sinks and troughs. Other varieties are worth experimenting with in the open. Plant them in full sun where the soil can be kept moist at all times, or else in gently shaded spots – the north-facing side of banks, shady sink and trough gardens and crevices between boulders on rock beds. They are happy, too, wedged into lumps of tufa – though you'll have to make sure that their pads are not overrun with moss. In pots water the

plants freely when necessary from early spring to autumn, but keep them very much on the dry side in winter. Those planted outdoors should be sited in well-drained soil (preferably enriched with a few limestone chippings) which is not likely to be soggy in winter, but which never dries out in summer – the perfect compromise! Repot the container-grown plants only when they have outgrown their pots, but give them a little top-dressing after flowering each year. Keep them plunged in a shaded frame in summer.
Encrusted saxifrages are not demanding at all – growing well in the rock garden in any well-drained soil that has been laced with a little peat or leafmould and a good helping of chippings, preferably of the limestone variety. If these are difficult to locate in your locality, use ordinary stone chippings and add a little ground limestone to the soil. The plants look and feel most at home in crevices – either in the retaining walls of rock beds or banks or between boulders on the rock garden. They'll also grow well in sink gardens and lumps of tufa. Most are happy in sun, but the large rosettes of *S. longifolia* and *S.* 'Tumbling Waters' seem to enjoy a shady, north-facing spot on my rock bank.

The others can best be dealt with individually. *S. oppositifolia* is happiest in a shady part of the rock garden in soil that is well drained yet never likely to dry out. It appreciates a little peat or leafmould and a good helping of limestone chippings. Kept well supplied with moisture it will look good and spread well, especially down a slope. It can be grown in pans as recommended for Kabschia saxifrages, but take care to keep it moist and shaded, or parts of its mat will turn brown.

S. × *urbium* will grow almost anywhere and in any soil in sun or light shade. The miniature forms are good in trough and sink gardens.

S. fortunei is grown to perfection in the peat garden, where it enjoys the acid conditions, the deep, moist root run and gentle shade. Create a special shady pocket for it on the rock garden if you like, making sure that it is sheltered from winds and so protected from the worst of the winter weather.

Propagation Most saxifrages can be propagated by division of the clumps immediately after flowering. Single rosettes of the Kabschias and encrusted saxifrages can be taken as cuttings in late spring and rooted in a propagator. Seeds of many species can be sown in pots of seed compost in early spring, and germinated in a garden frame.

Scabiosa (Dipsacaceae) Scabious, Pincushion Flower

The alpine scabious, *S. alpina* (also offered as *S. columbaria*), is a little European charmer with typical, pale-blue scabious flowers on 6-in (15-cm) stalks held over deeply slashed leaves in late spring and summer. At least, that's what you'll get from a nurseryman – the name is strictly out of date now, and its synonym (*Cephalaria alpina*) applies to a whopping plant with pale-yellow flowers. Never mind; the plant you buy is almost certain to be dwarf – especially if it is labelled *S. alpina* 'Nana'.

Cultivation A sunny spot in any ordinary, well-drained soil suits this alpine, and it especially enjoys chalky ground. Plant it in bright pockets on the rock garden, in rock beds and retaining walls.

Propagation Seeds can be sown as soon as ripe in pots of seed compost; germinate them in a garden frame. Divide clumps in spring.

Scutellaria (Labiatae) Skullcap

You're never likely to rave wildly about the skullcaps, but they do provide a bit of welcome colour in summer, and so are worth growing to extend the season of interest of any rock feature. *S. alpina* (Europe and Russia) is a spreading plant that grows to about 6 or 8in (15 or 20cm) and carries hooded purple flowers with a white lower lip. *S. scordiifolia* (Korea) is the same size, and has blooms of rich, violet-blue.

Cultivation The one thing about the skullcaps is that they are undemanding. Give them any sunny spot in a patch of well-drained soil and they'll thrive – tossing their stems down banks, over walls and rock faces. Keep them away from slow-growing alpines, which they may smother.

Propagation Seeds can be sown in pots of seed compost as soon as ripe, and germinated in a garden frame. Cuttings can be inserted in pots of sand/peat mixture and rooted in a frame in summer.

Sedum (Crassulaceae) Stonecrop

'This vast race, as a whole, is curiously uninteresting', says Reginald Farrer. 'I can think of no crime – not even a little one – that I would commit to obtain a Sedum', says Clarence Elliott. So here am I to spring to the rescue of these pleasing little plants. But 'little' is really the wrong word, for it's their rampant and bullying antics that have caused them to fall from grace with growers of choice alpines. Now and again, though, there's a place for a plant that sheets the ground with a neat array of succulent foliage and starry flowers, caring little for the soil and situation or for any solid obstacles put in its way. But I have to admit that there are sedums to avoid. *Never* let *S. acre* loose on your rock garden. You can let it loose on patios, by paths and anywhere else that is devoid of plants, but set it free on the rock garden and it will become a rival to oxalis and marestail – springing up in every nook and cranny. Of the 500 or more different sedums I offer my choice of just eight! Most are of similar culture, so you can follow my notes for any others you can lay your trowel on. *S. cauticolum* (Japan) and *S. sieboldii* (Japan) are similar to one another, but the first has crimson flowers and greyish leaves carried in threes, while the second is pink flowered, and its greyish leaves are paired. Both flower in autumn, both die down in winter, and both produce the most superb wine-red tints before they disappear. In summer there is no brighter sedum than *S. kamtschaticum* 'Variegatum'. The leaves are spoon shaped, green edged with cream, and the flowers are orange and red. The leaves, like those of the two previous species, are burnished in autumn. *S. middendorffianum* (Manchuria) produces a clump of upturned stems, 4 to 6in (10 to 15cm) high, with narrow, plum-purple leaves. Yellow flower clusters open in midsummer. One of the most beautiful of all sedums (and one I'd consider committing a small crime for) is *S. pulchellum* (North America). It's a sprawler with tiny, inflated green leaves and in summer it produces 6-in (15-cm) stems carrying heads of pink flowers that are branched so that they look just like starfishes. *S. rhodiola* (syn. *Rhodiola rosea*) is the British native rose root, whose roots are supposed to smell of roses. I've just excavated a hole to test the theory and can confirm that they are sweetly scented – though to say they are rose scented is rather too romantic. The 9-in (23-cm) stems are clothed in spirals of grey leaves and topped with clusters of yellow-green flowers in spring. They die down in autumn. Most gardeners know *S. spathulifolium* (North America) in its form 'Cappa Blanca', with purple-tinged leaves that are covered in white farina. Its blooms are yellow. *S. spurium* (Caucasus) is my final choice, preferably in one of its deep-red-flowered varieties such as 'Erdblut'. It blooms in early to mid-summer.

Cultivation In general the sedums like bright, sunny spots in poor, well-drained soil. Plant them in the crevices between rocks and paving stones and in scree or rock beds. They also do well when allowed to cascade over walls. Most are too rampant for sink gardens.

S. cauticolum grows best for a friend of mine in a peat bed, and *S. pulchellum* also appreciates a little shade and a little more moisture at the roots. All are happy in chalky soil as a rule, and are useful for brightening up alpine lawns in summer and autumn. Dwarf bulbs can grow easily through the carpeting types. The sedums that die down during winter are best planted where their absence will not be too obvious. The plants are sometimes grown in pots to add brightness to the alpine house in summer and autumn. Pot them up in a mixture of equal parts John Innes No. 1 potting compost and sharp grit, and water them freely when necessary from spring to late summer, much more sparingly in autumn and winter, keeping the compost almost bone dry. Occasional liquid feeds are appreciated in summer. Repot after flowering when the plant has outgrown its container.

Propagation Shoots and single leaves can be taken as cuttings in summer, and rooted in a peat/sand mixture in a propagator. Seeds can be sown in spring in pots of seed compost, and germinated in a warm propagator.

Sempervivum (Crassulaceae) Houseleek

It's always amusing to see the alpinist's reaction to sempervivums, for they are in a similar category to pleiones. The pleione is difficult to like because it's an orchid, and the sempervivum is shunned because it's almost a cactus! But plant your sempervivums sensibly and they are invaluable – offering variety of form right through the year, bright flowers in summer, and often rich, autumn tints. The genus is as bulky as *Sedum*, comprising masses of species and varieties. The enthusiast would do well to get hold of a little booklet published by Alan Smith of 127 Leaves Green Road, Keston, Kent, whose nursery stocks an unrivalled collection of sempervivums and their near relations the jovibarbas. I'll excuse myself from reciting reams of names – you might just as well pick the ones you fancy the look of in the nursery – but three I can vouch for are *S. arachnoideum* (European Alps), the cobweb houseleek, with its tiny rosettes clad in spider's-web silk. It pushes up lovely clusters of starry, pink flowers in summer, and spreads into neat but sizeable pads. *S. tectorum* (Europe) is the common houseleek, with large rosettes that are green, tipped with plum purple. The flowers are various shades of pink, and the plant has produced plenty of hybrids, some with rosettes that are burgundy red. *S. guiseppii* (Spain) makes a pleasant change from most by being totally covered in tiny, white hairs that give it a pinkish tinge. Its leaves are also tipped with deep purple. Dozens of hybrids will tempt you –

many of them attractively tipped and flushed with bright red.

Cultivation The plant's common name comes from the age-old custom of planting it in dollops of clay on house roofs to repel lightning. I cannot vouch for its efficacy, but it does brighten up both walls and rooftops when so grown. Otherwise plant it in full sun in any well-drained soil so that it can spread through rock crevices, chinks in paving and in vertical crevices in dry walls. Insert young plants as building progresses. The rosettes make good eye-catchers in alpine lawns, and are a gift to the sink gardener who can plant twenty or so varieties in a good-sized trough. Beware of the giants though. One monster is already taking over a sink of mine, and it's only been there for six months. Remember that some of the plants are neater growing than others – *S. arachnoideum* is restraint personified compared with some species. In pans in the alpine house sempervivums are a boon. Plant them in a mixture of equal parts John Innes No. 1 potting compost and sharp grit, and water them freely as soon as they are dry in summer. In winter apply water only when the rosettes start to shrivel slightly. Stand the pans outdoors in summer in full sun, and repot them in spring when necessary. Individual rosettes will die after flowering. Vast panfuls can be given monthly liquid feeds in summer.

Propagation Remove and pot up individual rosettes in spring or summer, transferring them to the rock garden when they are well rooted.

Shortia (Diapensiaceae)

Mouthwateringly beautiful, evergreen woodlanders that are often tricky to grow, these are always expensive to buy – if you can locate them. One or two nurserymen do carry small stocks, so I feel I can legitimately include them. *S. galacifolia* is the most difficult to obtain. It's a North American plant which creeps slightly, sending up rounded, glossy, leathery leaves of rich green, tinted with red in autumn. The single, toothed, blush-pink or white bell-flowers are carried nodding above the foliage on 6-in (15-cm) stems in spring. *S. soldanelloides* (syn. *Schizocodon soldanelloides*) is Japanese, and has larger leaves than the previous species and bells that are deeply fringed – just like a soldanella. The flowers are more generously carried here, too, with half a dozen or more on a stem, and they are tinted a deeper shade of pink. *S. soldanelloides* 'Magna' is altogether larger; *S. soldanelloides* 'Alpina' is neater, and *S. soldanelloides* 'Illicifolia' has indented leaves that give them a holly-like appearance. *S. uniflora* has a touch of class about it. It's Japanese and sometimes known as Nippon Bells (which adds nothing to its image). It makes a low, slow-creeping mat from which rise leathery leaves that are toothed at the edges and often red tinged. The flowers appear singly on 6-in (15-cm) stems in spring, and are glorious bells with scalloped and indented rims displaying a cluster of stamens in the centre. The blooms are blush pink and at their largest and most coveted in the form 'Grandiflora'.

Cultivation Without doubt the best and most long-lived shortias are grown in peat beds where they are shaded from sun, equipped with organic, lime-free soil and an uninterrupted supply of moisture. Northern gardeners can always grow better plants than southerners, for their climate is cooler and usually wetter. Dry summers will see shortias off in no time, and drying winds will burn them to death. As well as thriving in the top of a peat bed, the plants will also flourish in walls of peat blocks, provided these are never allowed to dry out. In pots the plants can be grown in a mixture of equal parts John Innes ericaceous compost, sieved leafmould and coarse peat. Keep them plunged in a shaded frame all the year round except when they are in flower, when they can be brought into a shaded part of the alpine house. Keep the compost gently moist at all times – even in winter – using rainwater rather than tapwater. Limy water will bring about their demise. Spray them with rainwater daily in hot spells. Repot after flowering when the existing container is outgrown. Annual top-dressing with the potting mixture is advisable in early spring.

Propagation This is difficult. Division of mature plants can be carried out in early spring, and the youngsters potted up individually in the recommended compost plus an equal amount of sharp sand. Keep the plants shaded and moist to aid establishment, and plant them out in autumn. Sometimes divisions refuse to establish themselves. Seeds can be sown in lime-free seed compost in spring, but the plants are slow growing and may well tire of trying to reach flowering size. This is undoubtedly one of those plants for the accomplished or dogged grower!

Silene (Caryophyllaceae) Catchfly

Bright relations of the British native red campion, these are fine plants for rock gardens, bringing summer brightness with their rosy-pink or white flowers. *S. acaulis* (Europe, including Britain) is the moss campion, a tufted mat-former that smothers itself in stemless, pink catchfly blooms in late spring and early summer – if it is happy. If it's not it won't bloom. *S. schafta* (Caucasus) i much more reliable, offering masses of pinkish-purple flowers on 6-in (15-cm) stems in mid to late summer. It spreads sideways in an orderly fashion. *S. keiskii* 'Minor' is perhaps worth a try, but it's always disappointed me. Its narrow leaves are dark plum purple when young, later becoming glossy green, and its flowers rich pink. It grows to around 4in (10cm) with me, flowers off and on in July and August, and makes rather a scrappy plant. Perhaps I should persevere with it – I've only had it a couple of years and it may improve with age. *S. maritima* 'Flore Pleno' is the double-flowered version of the British and European bladder campion. It is rather more orderly than the true species with similar narrow leaves, but its blooms are rather like those of the pink 'Mrs Sinkins'. The calyx is inflated, and the white petals are carried in profusion. It is an interesting little plant.

Cultivation To make *S. acaulis* flower well you'll have to grow it in a scree bed or in a gritty, impoverished pocket of soil on the rock garden. Plenty of sun and a lean diet suit it best. The others will do well in any sharply drained soil (chalky or not) and full sun, throwing their stems over rocks and down sunny slopes with abandon.

Propagation Seeds can be sown in pots of seed compost in spring, and germinated in a garden frame. The plants are not always long lived, and a few young replacements should be grown on to replace your elderly residents. *S. acaulis* can be divided in summer, as can *S. schafta*, which can also be propagated from summer cuttings rooted in a frame containing a mixture of peat and sand.

Sisyrinchium (Iridaceae) Satin Flower

These are neat rock-garden plants with tufts of miniature 'iris' leaves and starry flowers. *S. angustifolium* is, like all the species I'll mention, from North America, but it has become naturalised in parts of Britain. Its common name is blue-eyed grass, for it produces small clusters of blue, six-petalled flowers at the tips of its 6-in (15-cm) stems in summer. It will sometimes seed itself round a little too freely. *S. brachypus* is similar, just a little more robust and yellow flowered. *S. douglasii* (syn. *S. grandiflorum*) is the spring satin flower, and grows rather taller than the other two at 9 or 12in (23 or 30cm). It's a beauty. The leaves are narrow and rush-like, but the flowers are open bells of iridescent reddish purple on slender, almost nodding, stalks. Unlike the other two species, this one dies down as soon as it has flowered in spring.

Cultivation The first two species – *S. angustifolium* and *S. brachy-*

pus are easy in any ordinary, well-drained soil that has been enriched with a little peat and sharp grit. They both like sun, but prefer a spot where the earth is not likely to become baked. Try them in bright pockets of the rock garden that receive direct sun for only part of the day. *S. douglasii* definitely prefers to be kept out of scorching sun, and is best positioned in dappled shade. Again, it appreciates a soil that has been well laced with peat and grit. It doesn't like chalky soil. Mark it well when it dies down after flowering. It is often grown in the alpine house, where its early blooms can be appreciated and protected from foul weather, but it seems to give its growers more trouble under these circumstances than when it is planted in the open. In pots grow it in a compost consisting of equal parts John Innes No. 1 potting compost, peat and sharp grit. Water very carefully at all times, keeping the compost gently moist while growth is active, and barely moist when the plant dies down. Sogginess of the compost will lead to rotting. Shade the plant from bright sunshine. Repot after flowering when necessary.

Propagation Divide the plants carefully in late spring. Self-sown seedlings of *S. angustifolium* can be lifted and transplanted in spring.

Soldanella (Primulaceae) Blue Moonwort; Snowbell

I first glimpsed this plant in the Bavarian Alps. After a long climb up rough and rocky paths we came to a hidden valley where the snows were slowly receding from short, sheep-mown grass. The thick, cold mist suddenly lifted, and there, in the sodden lawn, nestled the dainty, fringed bells of *S. pusilla*, fresh as daisies and almost as numerous. Enough of the romance! There are several species that are easily obtainable, and are best grown in sinks and troughs or in pots in the alpine house – somewhere where you can keep your eye on them. All bloom in early spring, and all make neat clumps of rounded, leathery leaves. *S. alpina* (Eastern Alps and Pyrenees) has deeply fringed bells of lavender blue carried in clusters on 3-in (8-cm) stalks. *S. villosa* (Pyrenees) is similar but rather coarser, and grows to 9in (23cm). *S. montana* (Europe) is not quite so beefy at 6in (15cm) high, nor so hairy. Grow any species you are offered; they're all delightful.

Cultivation You can grow soldanellas in the open rock garden if you like – finding for them a sheltered, cool and gently shaded spot where the soil has been generously enriched with grit and plenty of peat. They enjoy chalky soil so long as plenty of organic enrichment has been added. However, I think you'll have more success

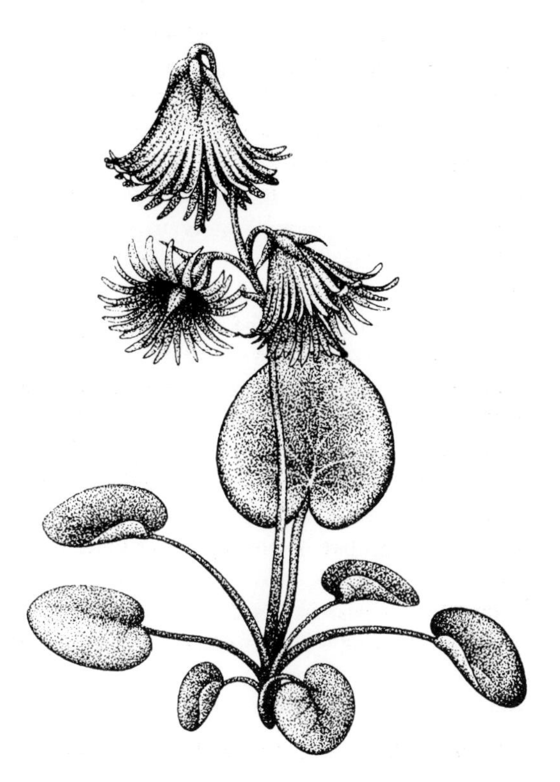

Figure 2.9
Soldanella alpina
Alpine Snowbell

at flowering them where they are grown in sinks, troughs and pots. Slugs love the overwintering flower buds, and often demolish them before the blooms open. In a sink containing a mixture of John Innes No. 1 potting compost plus an equal amount of sharp grit the plants will do well, and can be protected from slugs if mini-pellets are scattered among the leaves every few weeks during autumn and spring. Keep the trough in a shaded spot and, if possible, protect it from excessive rain in winter (place a cloche over it if you can). This covering will also help to prevent the flower buds from damping off – another common cause of failure – and from severe frosting. In pots the plants grow well in a mixture of equal parts peat, sharp grit and John Innes No. 1 potting compost. Slugs are less likely to be a problem here, but the plants may still be reluctant to flower if you are mean to them. Keep them in a decent-sized pot, repotting when necessary after flowering. Water freely in spring and summer to keep the compost nicely moist, but keep it slightly on the dry side in winter. Frequent division of the plants (when they are two or three years old) helps to rejuvenate them. Old plants may not be so free flowering.

Propagation This can be done by division of the clumps immediately after flowering. Seeds can be sown as soon as ripe in pots of seed compost plunged in a shady frame. Germination may be slow.

Synthyris (Scrophulariaceae)

They're never likely to set the world on fire with their brilliance, but these plants do bring variety to your spring flower collection. There are two species generally available: *S. stellata* (north-western America) is herbaceous and has lobed, hairy leaves which push up in an orderly clump with the spires of violet-purple flowers in March and April. *S. reniformis* (Washington, Oregon, California) has lobed, kidney-shaped, evergreen leaves and rather more open clusters of tubular, lavender-blue flowers on 4 to 6-in (10 to 15-cm) stems.

Cultivation Both plants need a moisture-retentive soil that is rich in leafmould and peat, but which is unlikely to remain soggy in winter. They do best in shade, but will put up with sun if they have moisture at the roots. Avoid planting them in chalky ground. They do best in the peat bed or a well-prepared shady pocket on a rock bank. Top-dressing with leafmould in spring is beneficial.

Propagation Divide mature clumps immediately after flowering.

T

Tanacetum (Compositae)

The description of *T. densum amanum* (syn. *T. haradjanii*, *Chrysanthemum haradjanii*) that I once read somewhere struck me as so accurate that I've never been able to think of its leaves as like anything other than 'deeply cut jerseycloth'. The plant is a prince among silver-foliage carpeters, making dense, 4-in (10-cm)-high rugs of its greyish, feathery leaves that in summer may be studded with tiny, yellow flowers which some cynics liken to groundsel. Snip them off if you don't like them. The plant comes from Turkey and Syria.

Cultivation Bright sunshine and a well-drained soil is all the plant asks for. Site it at the edge of a rock bed where it can run over the rim, or at the foot of a rock garden – provided it will not sit in water during the winter. Don't grow it in a pot – it's such a beauty when allowed to roam for 3 or 4ft (1m or 1m 25cm). Brown or bare patches can be snipped out with a pair of secateurs (though they are only likely to occur on older plants which are best replaced every three years or so).

Propagation The plant can be layered in summer, or stem cuttings

2in (5cm) long rooted in a propagator in summer. Dip them in fungicide to prevent rotting.

Teucrium (Labiatae) Germander

The teucriums are useful little shrublets for bringing colour to late summer – often a lean time on the rock garden and in the alpine house. Nurserymen offer a surprising range of species, all of which are worth a try. *T. aroanium* (Greece) grows only a couple of inches (5cm) high but makes a mat a good foot (30cm) across. The undersides of the leaves are white and woolly and the hooded flowers lavender-blue veined with purple. *T. polium* (syn. *T. aureum*) (southern Europe and south-west Asia) makes a thick mat of toothed grey leaves decorated with whitish or yellow flower clusters. It may rise up to as much as 1ft (30cm) on occasions. *T. chamaedrys* (Europe), the wall germander, is the beefiest of the bunch, growing 1ft (30cm) high and more across. Its leaves are green and its purplish-mauve flowers carried in tall spires. For variety of foliage you should certainly grow *T. scorodonia* 'Crispum' (often listed as *T. scordium* 'Crispum') – a crimped-leaved variety of the British native water germander. Its crested, green leaves are a treat to look at – it's as if they'd been gathered up around the edges to make a frill. They may be tinged with purple in winter, and the plant grows upwards to about 1ft (30cm). Finally, give *T. subspinosum* (Balearic Islands) a go. It makes a thick, squat little shrublet clad in leaves that are green above and grey beneath, and its stems may end in spines. The little hooded flowers are pinkish lilac, and like all the other species open in mid to late summer. In common with many of its relatives this plant is aromatic, so keep your cat away from it!

Cultivation Sun-worshippers all, these are plants for bright screes or well-drained spots high up on the rock garden where they can spread their mats and bask in the warmth. *T. aroanium* is good in sinks and troughs, and others can be planted in rock beds and atop walls so that their mats tumble over the edges. The plants can also be grown in an alpine house or frame in pots containing a mixture of equal parts John Innes No. 1 potting compost and sharp grit. Water the plants freely according to demand in spring and summer, but keep the compost very much on the dry side in winter. Repot annually in spring.

Propagation Cuttings of firm shoots can be rooted in a propagator in summer.

Thalictrum (Ranunculaceae) Meadow Rue

More alpine gardeners should grow the Japanese *T. kiusianum*. It rises only to around 6in (15cm), and produces grey, maidenhair-fern-type foliage over which are carried fluffy, pink flowers on fusewire stems. No plant is daintier. It spreads itself quite respectably, so it's never a real nuisance, and in early summer it looks a treat – a subtle symphony of pink and blue-grey.

Cultivation In the open the plant likes an ideal soil – a mixture of earth, peat, leafmould and grit. This provides it with the moisture-retentive yet well-drained medium that is most to its liking. It also needs gentle shade – those fragile leaves are thin and will scorch in burning sun. If your garden is chalky, work in plenty of organic matter and the plant should survive. Shady pockets on the rock garden will suit it, but it can also be grown in pots in an alpine house or frame. Pot it up in a mixture of equal parts John Innes No. 1 potting compost, sharp grit and peat. Apply water freely as necessary while the plant is in growth, but keep the compost on the dry side once it has died down. Repot in spring when the plant has outgrown its container.

Propagation Divide large plants in early spring.

Thymus (Labiatae) Thyme

The scent of thyme on a warm summer's evening, coupled with the misty, lilac-pink or the deep purple of its flowers is a combination that assures it of a place in any sensitive gardener's patch. There are dozens of different species and dozens more different varieties, most of which are reliable and floriferous. It should come as no surprise to you that the nomenclature of the thymes is confused, but if you can buy a plant whose habit you approve of and whose flowers you admire, it matters little what it should be called. Some you might appreciate are those sold as: *T. serpyllum* 'Minus', which makes flat, downy, spreading pads of growth smothered in pink flowers in summer; *T. serpyllym* 'Bressingham' is quicker spreading, up to 3in (8cm) or so high, and carries rich lilac-pink flowers; *T.* 'Silver Posie' has leaves variegated with creamy white, and *T. citriodorus* 'Aureus' grows to 6 in (15cm), is yellow leaved and lemon scented. If you want a challenge, try *T. cilicicus* (Asia Minor), which makes a little upturned plant to 3in (8cm) high. It has downy leaves and in summer thick 'bottle-brush' flowers of pink at the tips of its stems. *T. membranaceus* (Spain) is another tricky customer that makes a dome of a bush up to 6 or 8in (15 or 23cm), hopefully crowded in early summer with white flowers nestling among conspicuous, pink, papery bracts. Well-grown plants may be 1½ft (45cm) across – what a sight!

Cultivation Most thymes are blissfully happy when given a well-drained patch of soil (chalky or not) with their heads in brilliant sunshine. You'll get the full impact of their aroma if you grow them between paving stones on a patio where your feet and deck-chair legs will occasionally crush the leaves. On scree beds, in sink and trough gardens they are also good, and really flat ground-huggers like *T. serpyllum* 'Minus' make superb 'kerbs' for the rock garden or bank. The two tricky species, *T. cilicicus* and *T. membranaceus* can be grown outdoors in very light soil and full sun if they can be sheltered from the worst of the winter weather – with a cloche or a pane of glass, but they are more reliably cultivated in pots or pans in the alpine house or frame. *T. membranaceus* has a habit of dying off for no apparent reason when it has made a good-sized plant, though this is probably due to damp weather rather than severe cold. Some gardeners speak of it as dying regularly in its second or third year, but plants have been grown to the age of at least eight – flowering well, too. If you grow these two in pots, plant them in a mixture of equal parts John Innes No. 1 potting compost and sharp grit. Water them freely when they ask for it between spring and late summer, but then let them remain much drier (though not dusty) until the following spring. Keep them in a sunny, well-lit spot at all times. Repot every year in spring. All species and varieties can be clipped free of faded flowers when they start to look unsightly.

Propagation Cuttings can be rooted in a propagator in early summer, and layered portions detached and replanted at the same time. Seeds can be sown in pots of seed compost in spring, and germinated in a garden frame or propagator.

Trillium (Liliaceae) Wood Lily

Without doubt the wood lilies are an acquired taste, but once acquired they'll continue to captivate you annually with their trio of leaves and three-petalled flowers. All flower from early to late spring, and all those mentioned here come from the USA. There are a dozen or so different trilliums relatively common in cultivation, and most can be picked up from one source or another. By far the commonest and I think one of the most beautiful, is *T. grandiflorum*, the wake robin. The trio of deep-veined, rich green leaves rises on a stalk up to 1ft (30cm) high, and from the centre of the foliage springs a 2-in (5-cm)-wide, three-petalled flower of pure white, blushing a little as it ages. The rare double form is one of those plants that can honestly be described as breathtaking, and fortunately it is now available in small quantities each year from at least one nurseryman. You'll have to pay dearly for the pleasure of owning it, but it's not difficult to grow, only slow to increase. It seems insulting to describe the luscious, white flowers as being like double daffodils without a trumpet; perhaps it sounds more romantic if they are likened to gardenias. Quite different is *T. sessile* – its three leaves are marbled with bronzy brown, and the three petals in their centres are maroon, strap shaped and upturned to make a sort of cockade. I think it's one of the most spectacular, if not the most typical of wood lilies. *T. erectum* is vulgarly called lamb's quarters. Here the flowers are three-petalled and starry and coloured rich maroon with a central splash of golden stamens. The plant has always had a reputation for smelling foul, but you'll have to put your nose right into the flower before you can detect what is certainly an unpleasant odour. Maybe that's how it came by its common name!

Cultivation The trilliums are not difficult plants if you give them the soil and conditions they love best. They need a shady spot, even overhung by trees, and a soil which is rich in humus. The peat bed is without doubt the place where the best plants are grown, but in a shady spot and a good soil that has been generously enriched with leafmould and peat they'll thrive, provided the earth never dries out. Each spring a good top-dressing of well-rotted leafmould or peat will continue to keep the soil in good shape. If you garden on chalky soil don't despair totally of growing trilliums. Really prepare your soil well, digging in as much organic stuff as you can, and then try a single species such as *T. grandiflorum*. It may grow well. Some gardeners with alkaline soil have difficulty with trilliums, others manage nicely. By the way – do be careful not to disturb your plants once they have died down. Their pointed buds are not easy to see, but are very easy to damage with a wayward foot or fork. Mark their positions with canes if necessary, and try not to walk on their patch of soil at any time.

Propagation Clumps can be divided in autumn, and portions of rhizome with a bud can be transplanted. Seeds can be sown in March in peat-based seed compost with sharp, lime-free sand added. Germinate them in a cool, shaded frame and be patient; it will probably be two years before the shepherd's crooks show through the compost, and then they'll not emerge all at the same time. Don't prick them out individually; instead pot on the entire ball of compost into a larger pot containing a mixture of John Innes ericaceous compost, sharp grit and peat in equal parts. Transplant them later during the dormant season.

Tropaeolum (Tropaeolaceae)

In nature colours never clash, they say. It's a statement I've never believed, for nature has more basic intentions at heart than providing William Morris effects for the benefit of man. But every now and again you'll come across a plant whose colour combination is so cleverly calculated that it seems to have been contrived by a masterful artist. Who in their right mind would dream of putting together bright yellow and blue-grey? That's just the combination to be found in *T. polyphyllum*, a Chilean plant with 3-ft (90-cm)-long trailing stems thickly clothed in five-lobed glaucous leaves and then masses of spurred, yellow trumpet flowers in June or July. It's a masterpiece.

Cultivation You'll buy this plant in its dormant form – as a tuber looking something like a fir-apple potato. Plant it 1ft (30cm) deep in a well-drained soil that's been well laced with peat and grit. It insists on bright sunshine, and should be positioned where it can cast its stems down a bank or over a retaining wall. It dies down after flowering and re-emerges the following spring. Be warned – it may not always appear in the same spot every year!

Propagation It's best not to disturb the plant once it's established, but the tubers (if you can dig deep enough to find them) can be separated and transplanted in autumn.

Uvularia (Liliaceae) Bellwort

Demure little woodlanders with arching stems clad in curling Solomon's-seal-like leaves, and tipped with nodding, pale-yellow flowers with twisted petals, the bellworts have a quiet charm that endears them to many folk. *U. grandiflora* (USA) is the type most freely available, and it grows to about 9in (23cm), flowering in May.

Cultivation Shady spots in peaty pockets, or in a peat bed is where these plants grow best. Make sure that the soil never dries out and that scorching sun never reaches the plants. Top-dress annually in spring with peat or leafmould.

Propagation Divide clumps in spring just as growth is starting.

Verbascum (Scrophulariaceae) Mullein

One verbascum above all others is now grown by alpine enthusiasts in rock gardens, frames and alpine houses. It's *V.* 'Letitia', spotted at Wisley by the late Ken Aslet and named after his wife. It was a chance seedling which had as its parents *V. dumulosum* and *V. spinosum*, and it makes a dome-shaped bush around 6in (15cm) high and 1ft (30cm) across. The neat spires of clear-yellow blooms are stained at the base with rusty orange, and centred by a tuft of red-stalked, orange-tipped stamens. They open throughout the summer. The leaves are oval, slightly indented along their edges, and felted with a thick, grey down. Altogether it's a little charmer.

Cultivation In the rock garden the plant needs a sunny spot in a gritty, free-draining soil. It doesn't mind chalk, and especially appreciates scree conditions, though it's also useful in sink and trough gardens. In the alpine house grow it in pots containing a mixture of equal parts John Innes No. 1 potting compost and sharp grit. Water freely as necessary in spring and summer, but keep the compost on the dry side through the winter. Repot every year in spring, and cut out any dead shoots as soon as they are seen. Faded flowerheads should be snipped off, too.

Propagation Root cuttings can be removed in spring, trimmed to a length of 2in (5cm), and inserted vertically in a mixture of peat and sand in a frame. Pot them up individually when each shoot is well established.

Veronica (Scrophulariaceae) Speedwell

Here's a group of plants guaranteed to make the fastidious gardener throw in the towel. Half the plants that used to be veronicas are now called *Hebe* or *Parahebe*. Those that remain in the genus have all been 'sorted out', so that the names you once knew them by may not be their true names at all. What's a writer to do when confronted with such a muddle? Give all the new names, knowing that the reader won't be able to find the plants in nursery catalogues? Give all the synonyms and fill five pages and the reader's mind with confusion? Or take up an ostrich-like attitude and ignore the new names, giving just those that are likely to be found in catalogues? On this occasion I'll plump for the ostrich, and risk the upshot. *V. incana* (Europe and Asia) is one of my first choices. It makes a rug of narrow, white-felted leaves from which rich-blue, 9-in (23-cm) flower spikes shoot up in summer. *V. gentianoides* (Caucasus, Crimea) has 1½-ft (45-cm) spikes of larger flowers up to ½in (13mm) across; each one pale-blue veined with a darker shade. The foliage is rich green, except in the variety 'Variegata', in which it is splashed with white. For smaller rock gardens *V. prostrata* (Europe), in one of its many varieties, will be more suitable. It grows to just 6in (15cm), and smothers itself in summer with flower spikes of light blue, dark blue or pink. *V. cinerea* (Turkey) can be fitted in anywhere, too. It makes a ground-hugging mat of greyish, evergreen leaves that are a fine foil for the early summer flowers of bright blue.

Cultivation These are very obliging plants, flourishing in most well-drained soils, chalky or otherwise. Give them a sunny spot. The dwarf kinds look particularly good when used to edge rock beds or rock banks or even flower borders, and they provide useful vertical colour where a lift is needed for flat plantings. After several years the plants may become scrappy, at which time they should be divided up and replanted.

Propagation Divide mature clumps in spring or autumn, and replant healthy young portions.

Viola (Violaceae) Violet, Pansy

Part of romantic folklore, the violets really earn their keep in woodland and on rock gardens, where they will bloom for an amazing length of time in spring and summer. I offer you a brief selection of personal favourites that are easy and rewarding. *V. cornuta* (Pyrenees) is the horned violet, a far from tidy plant, but one which blooms as long as any – right through the summer. Its spurred blooms are carried at the top of 6-in (15-cm)-high, sprawling plants and are a good inch (25mm) across. There are dark-blue, light-blue and white forms – all are desirable – and the foliage is fresh green. If you like your plants to be as eye-catching in leaf as they are in flower, go for *V. labradorica* 'Purpurea'. Its plum-purple, heart-shaped leaves look a treat when mingled with the foliage of the acid-yellow creeping jenny (*Lysimachia nummularia* 'Aurea'), and the coy, violet-blue flowers nod on 4 to 6-in (10 to 15-cm) stems in early summer. *V. cucullata* 'Striata' (or 'Bicolor') has fresh-green leaves and white flowers veined with blue. A patch by the gate in my last garden cheered up passers-by every April. It makes spreading clumps (but does not run around quite so freely as *V. labradorica*), and blooms in spring. There are some superb hybrids to choose from. Take a chance with 'Bowles' Black', very dark blue, almost black; 'Penny Black', the blackest flower I've ever seen; 'Maggi Mott', pale blue, long flowering and sweetly scented; and 'Irish Moll' or 'Irish Molly' – a green and brown flower guaranteed to stop you in your tracks.

Cultivation All these plants appreciate shade from strong sunshine and a leafy soil that is constantly moist. Make pockets especially for them at the foot of shade-casting boulders, where they can luxuriate and sink their roots into the cool medium. If a stream runs through your rock garden site them up on the banks where the soil is not muddy, but never dries out. If all you can provide is a sunny spot, don't worry. Lace the earth with plenty of peat and leaf-mould, never let it dry out and the violas may still thrive and bloom well. Moisture is the secret of success.

Propagation Divide mature clumps and plantations in spring or autumn. Individual portions of 'runner' with roots can be planted at 6-in (15-cm) spacings to give ground cover. Seeds can be sown in pots of peat-based seed compost in spring, and germinated in a garden frame.

Waldsteinia (Rosaceae)

The glossy, strawberry-like leaves of *W. ternata* are evergreen, and spread thickly on the ground to make a 4-in (10-cm)-high mat which is studded in early summer with yellow 'strawberry' flowers. It's a useful carpeter that comes from Europe and eastern Asia.

Cultivation In sun or shade the plant will do very nicely, provided that the soil never dries out. Dusty earth will not please it one bit, so make sure that plenty of organic matter is worked in before you plant. Kept supplied with water, the thick mat of growth will suppress any weeds, and is particularly useful around shrubs where other plants are not keen to grow.

Propagation Divide the mat in spring or autumn, and replant portions at 6-in (15-cm) spacings to give rapid cover.

Wulfenia (Scrophulariaceae)

There's just one wulfenia offered with frequency by nurserymen and that's *W. carinthiaca* (eastern Alps and the Balkans). It makes clumps of thick, glossy, serrated leaves from which emerge thick spikes of violet-blue flowers in summer, and occasionally once more in autumn. The plant might reach 1ft (30cm) in height when happy and well grown.

Cultivation Give wulfenia a cool and shady pocket on the rock garden which has been enriched with peat, leafmould and grit. A good, full-bodied soil will produce good, full-bodied plants. On dry, sandy earth death will be only a hair's breadth away. Make sure that the soil never dries out from one end of the year to the other.

Propagation Divide mature clumps in spring.

Zauschneria (Onagraceae) Californian Fuchsia

This is one of those plants that needs a really good summer to do well, but it is such a spectacular sight when in bloom that it's worth the risk. *Z. californica* (California and Mexico) is the species most frequently grown. It makes foot-high (30-cm) stems clad in narrow, grey leaves, and blooms (weather permitting) in August for as long as the weather will allow – often into October or even November. The tubular flowers look more like penstemons than fuchsias, and are a rich, flowing scarlet. The form 'Mexicana' is reputed to come into flower earlier than the true species, and some nurserymen now sell a form known as 'Glasnevin', which is described as being more reliable. The trouble is that is grows to almost 1½ft (45cm) – making it difficult to fit into small rock gardens. Another species, *Z. cana* (syn. *Z. microphylla*) is found by some gardeners to be a better bet than *Z. californica*, and more likely to flower every year. Again it's rather tall at 1½ft (45cm). Don't let my warnings put you off giving at least one of them a try.

Cultivation As you'd guess from the foregoing, these plants need a hot, sunny position to force them into flower. Try them on a sunny scree bed or the top of a wall on a rock bed so that they can tumble down the sunny face. They will also grow from crevices in dry walls. They don't mind chalk, but they can't cope with faulty drainage, so work plenty of grit into the soil if you can't offer them a scree. Cut all the shoots back to about 6in (15cm) in spring to keep the plant in good shape and encourage the breaking of flower shoots from lower down the stems.

Propagation Cuttings of non-flowering shoot tips taken from quite low down on the plant can be rooted in a propagator in early summer. Seeds can be sown in pots of seed compost in spring, and germinated in a propagator. Old plants can sometimes be divided in late spring.

CHAPTER 3

Dwarf Conifers for Rock Gardens

I've omitted these entirely from the main text because they impinge only on the edge of the rock gardener's territory. Plant too many of them and you'll find that they rob the space needed for alpines, as well as giving your rock garden a funereal air. Used wisely and in moderation, a handful of conifers will add height and variety of form all the year round. Choose columnar and conical varieties where something is needed to lift the eye from a flat carpet of alpines, and use the rounded and prostrate kinds as evergreen backdrops for bright-flowered beauties.

The following list includes conifers that are well suited to rock gardens due to their slow rate of growth, habit and appearance.

Abies balsamea 'Hudsonia'
A rounded, green dome with fresh-green shoot tips in spring. Up to 2ft (60cm).

Chamaecyparis lawsoniana 'Gimbornii'
Rounded but rather upright with grey-green foliage held in close-packed fans. Up to 3ft (90cm).

Chamaecyparis lawsoniana 'Minima Aurea'
Broadly columnar in outline with densely packed yellow shoots. Up to 4ft (1m 25cm).

Chamaecyparis obtusa 'Nana'
A very slow-growing conifer with dark-green, curved fans of growth that are tipped with fresh green in spring. Up to 2ft (60cm). Good in sink gardens and pots.

Chamaecyparis pisifera 'Nana'
A low, tight-packed dome of green shoots. Around 1 or 2ft (30 or 60cm). Useful in sink gardens and alpine-house pots.

Cryptomeria japonica 'Vilmoriniana'
The dense mound of mid-green shoots has a slightly fluffy appearance. Up to 3ft (90cm).

Juniperus communis 'Compressa'
Justifiably popular, this juniper makes a tight column of grey-green up to 2ft (60cm) or more high, but very slowly. Ideal in sink gardens and in pots in the alpine house.

Juniperus horizontalis 'Glauca'
Useful where you want a spreading blue-grey carpet. Only 6in (15cm) high but up to 6ft (1m 85cm) across. Can be restricted by pruning in spring, or when unwanted growths get in the way.

Picea abies 'Gregoryana'
A very tight-packed hummock of fresh green. 1 to 2ft (30 to 60cm) high.

Picea glauca 'Albertiana Conica'
Neat pyramidal growth of mid-green with bright-green shoot tips in spring. Eventually 3ft (90cm) plus, but slowly.

Picea mariana 'Nana'
Makes a low dome of blue-green, tufted growth up to 1ft (30cm) high.

Pinus mugo 'Gnom'
Upturned pine stems with long needles make, in time, a rounded bush 3ft (90cm) or more high.

Thuja plicata 'Rogersii'
A dense orb of dark-green foliage burnished with gold. Up to 3ft (90cm).

All conifers are best planted in spring – then they'll grow away without any check. Plant from containers if you can, and work plenty of moist peat into the soil, plus a handful or two of bone-meal. Provided the soil is not allowed to dry out during the first year of establishment, the plants should soon settle in. If, in time, they outgrow their alloted space, cut around them with a spade about 1ft (30cm) from the stem, and lift them with as much soil around the roots as possible. With their roots wrapped in a piece of sacking or polythene they can be transported to a different part of the garden. Once again, spring is the best time to shift them, and copious watering will be necessary in dry weather. A spray over with a proprietary anti-desiccant fluid will prevent dehydration during the transplanting process.

Planted in pots, dwarf conifers can perk up the alpine house display at any time of year. I pot them in 6 to 9-in (15 to 23-cm) pots of John Innes No. 2 potting compost, which they seem to find very much to their liking. Repotting or top-dressing every year is advisable to keep them in good condition – do the job in spring.

CHAPTER 4

Dwarf Bulbs for Rock Gardens

One or two bulbous plants have sneaked into the main A-Z section because the gardener thinks of them as alpines rather than bulbs. But there are dozens more which I've not been able to squeeze in, so here's a list of some of the most reliable and beautiful rock garden bulbs that you are advised to find a home for in pockets of earth where they can bring a fleeting brightness to spring, summer, autumn or winter.

Name	Colour	Height	Season	Aspect	Plant
Allium karataviense	White	1ft (30cm)	May/Jun	Sun	Autumn
A. moly	Yellow	9in (23cm)	Jun/Jul	Sun/shade	Autumn
A. narcissiflorum	Pink	8in (20cm)	Jun/Jul	Sun/pots	Autumn
A. oreophyllum ostrowskianum	Mauve	6in (15cm)	Jun/Jul	Sun	Autumn
Chionodoxa luciliae	Blue/white	6in (15cm)	Mar/Apr	Sun	Autumn
Crocus ancyrensis	Orange	3in (8cm)	Feb/Mar	Sun/pots	Autumn
C. asturicus	Purple	3in (8cm)	Oct	Sun	L. summer
C. banaticus	Lilac	3in (8cm)	Sep/Nov	Sun	Summer
C. chrysanthus cvs	Various	3in (8cm)	Feb/Mar	Sun/pots	Autumn
C. etruscus	Lilac	3in (8cm)	Feb/Mar	Sun	Autumn
C. longiflorus	Lilac	3in (8cm)	Oct/Nov	Sun	L. summer
C. sativus	Purple	2in (5cm)	Sep/Oct	Sun	Summer
C. speciosus	Lilac	3in (8cm)	Sep/Oct	Sun/pots	Summer
C. tomasinianus	Purple	3in (8cm)	Jan/Feb	Sun/shade	Autumn
Eranthis hyemalis	Yellow	3in (8cm)	Feb/Mar	Sun/shade	In growth
Erythronium dens-canis	White/mauve	6in (15cm)	Mar/Apr	Shade	After flowering
E. revolutum	Pink	6in (15cm)	Apr/May	Shade	After flowering
E. tuolumnense	Yellow	1ft (30cm)	Apr/May	Shade	After flowering
Fritillaria meleagris	White/mauve	1ft (30cm)	Apr/May	Sun/shade	Autumn
Galanthus elwesii	White	9in (23cm)	Feb/Mar	Shade	After flowering
G. ikariae vars	White	6in (15cm)	Mar	Shade	After flowering
G. nivalis vars	White	4-6in (10-15cm)	Feb/Mar	Shade	After flowering

Dwarf Bulbs for Rock Gardens

Name	Colour	Height	Season	Aspect	Plant
Ipheion uniflorum	Pale blue	4in (10cm)	Mar/Apr	Sun	Autumn
Iris bucharica	Yellow	1½ft (45cm)	Apr/May	Sun/pots	When growing
I. danfordiae	Yellow	4in (10cm)	Jan/Feb	Sun/pots	Autumn
I. histrio cvs	Blue/white	4in (10cm)	Jan/Feb	Sun/pots	Autumn
I. histrioides cvs	Blue/white	4in (10cm)	Jan/Feb	Sun/pots	Autumn
I. reticulata cvs	Blue/purple	6in (15cm)	Feb	Sun	Autumn
I. winowgradowii	Yellow	4in (10cm)	Feb/Mar	Sun	Autumn
Leucojum aestivum	White	1ft (30cm)	May	Sun	In growth
L. autumnale	Blush	6in (15cm)	Sep/Oct	Sun	In growth
L. vernum	White	1ft (30cm)	Feb	Sun	In growth
Muscari armeniacum	Blue	6in (15cm)	Apr/May	Sun	Autumn
M. botryoides 'Album'	White	6in (15cm)	Apr/May	Sun	Autumn
Narcissus asturiensis	Yellow	3in (8cm)	Jan/Feb	Sun/pots	Autumn
N. bulbocodium	Yellow	4in (10cm)	Mar/Apr	Sun/pots	Autumn
N. cyclamineus	Yellow	6in (15cm)	Feb/Mar	Sun	Autumn
Ornithogalum balansae	White	4in (10cm)	Mar/Apr	Sun	Autumn
Puschkinia scilloides	Blue	4in (10cm)	Mar	Sun	Autumn
Scilla bifolia	Blue	6in (15cm)	Feb/Mar	Sun	Autumn
S. sibirica	Blue	6in (15cm)	Feb/Mar	Sun	Autumn
Sternbergia lutea	Yellow	6in (15cm)	Sep/Oct	Sun	Summer
Tecophilaea cyanocrocus	Blue	4in (10cm)	Mar/Apr	Pots	Autumn
Tulipa clusiana	White/red	1ft (30cm)	Apr	Sun	Autumn
T. kolpakowskiana	Yellow/red	8in (20cm)	Mar/Apr	Sun	Autumn
T. saxatilis	Lilac	1ft (30cm)	Mar/Apr	Sun	Autumn
T. tarda	Yellow/white	4in (10cm)	Apr/May	Sun	Autumn
Zephyranthes candida	White	6in (15cm)	Sep/Oct	Sun	Summer

As a general rule, bulbs and corms are planted to at least twice their depth. That is, a bulb which is 1in (25mm) from tip to base is planted in a hole 3in (8cm) deep, so that twice its own depth of soil sits on top of it. Those bulbs that are recommended for planting either shortly after flowering or while they are in growth, should be planted so that the soil mark on their fading stems is fractionally below the new soil level. These are bulbs which detest being allowed to dry out – buy them in the dry state and you are less likely to have a 100 per cent success rate.

Those bulbs recommended for planting in pots do not usually need to be planted so deeply. Just bury their 'noses'. I've found that most of them enjoy a mixture of equal parts John Innes No. 2 potting compost and sharp grit. Almost all bulbs will need repotting each year, just before growth commences.

Many alpine gardeners take to bulbs with enthusiasm, growing them both in pots and in the open. But there are some bulbs – from hotter climates – that need a baking after flowering. These can best be grown in a special 'bulb frame', where they can be given just the conditions they adore. *Bulbs Under Glass* by J. G. Elliott, published by the Alpine Garden Society, is essential reading for any enthusiast.

CHAPTER 5

Shrubs for Rock Gardens

Again I've cheated a little and included some dwarf shrubs in the A-Z of alpines because they are as much a part of any rock garden as gentians and aubrieta. Here I offer a few suggestions of additional shrubs that will add interest to any rock garden, bank or bed without overpowering the more legitimate residents.

Amelanchier alnifolia pumila
A 2-ft (60-cm)-high snowy mespilus with oval green leaves that make good autumn colour. White flowers open in May and are followed by black berries.

Berberis × *stenophylla* 'Corallina Compacta'
A tiny, evergreen barberry seldom more than 1ft (30cm) tall. The dark-green leaves show off well the tiny, orange flowers in May and June.

Calluna cvs
There are hundreds of heathers, valued for their bright bells in spring, autumn and winter, and their coloured foliage, which burnishes well in cold weather. Most grow no more than 1ft (30cm) tall, and a good many are ground hugging.

Daboecia cantabrica
A heath with large bell flowers and spiky foliage. Grows to around 9 or 12in (23 or 30cm) and is evergreen, making a thick mat of leaves. The bells are purplish pink.

Erica cvs
The heaths are too well known to need description, and are closely allied to calluna. Some species will tolerate chalk, unlike callunas. Height between 4 and 12in (10 and 30cm).

Gaultheria species
There are several low-growing and creeping gaultherias that are, in general, fond of a semi-shady spot in lime-free soil. They are evergreen, and usually produce ornamental flowers and fruits.

Hebe species and cvs
There are many hebes well suited to being grown on larger rock gardens, and one or two will be at home where space is restricted. Try 'Boughton Dome', which makes a conifer-like mound of rich, evergreen foliage up to 1ft (30cm) high. Other small species are valued for their flowers in shades of white, pink, or violet. Make sure you are not saddled with a rampant grower.

Hedera helix cvs
The ivies are fine plants, especially for covering the ground underneath shrubs which would otherwise remain as a seedbed for weeds. Choose the prostrate varieties that offer bright variegation or lustrous green colouring. Among the best are 'Glacier', white and green; 'Goldheart', green with a central yellow splash; and 'Sagittifolia' with finely fingered leaves. *H. helix* 'Erecta' is an upright-growing plant with close-set ranks of leaves, and it makes a handsome shrub up to 1ft (30cm) tall.

Kalmiopsis leachiana
Deep-pink, starry flowers are held over a foot (30-cm)-high evergreen bush in May. A choice shrub for the peat garden where it can sink its roots into cool, lime-free soil and bask in sun-dappled shade.

Margyricarpus setosus
Heather-like in appearance, this fuzzy, spreading shrub grows to around 4 or 6in (10 or 15cm), and carries white berries among its leaves from late summer to early winter.

Micromeria varia
A light-green, thyme-like shrublet with aromatic leaves and pale-pink flowers in June. It grows up to 6in (15cm) or so.

Rhododendron species
I dare not launch into a series of description, nor into a detailed account of cultivation. It's a superb genus full of breathtaking treasures. Find some of them a home in sun-dappled spots where there's always ample moisture at the roots in a peaty, leafy, generally lime-free soil. I bow my head in shame at being so brief.

Salix species
The dwarf willows are great little plants for the rock garden. Try *S. lanata*, with its woolly leaves carried on 1 to 2-ft (30 to 60-cm) stems, the prostrate *S. reticulata* with its glossy, deep-veined leaves, and *S. hastata* 'Wehrhahnii' where you've space to spare. It has purplish stems and green leaves, but its catkins are its main

attraction – silvery at first, turning sulphur yellow with pollen. Up to 3ft (90cm).

Sorbus reducta
Another plant for a bit of space. It will make a thicket of rowan-leaved stems about 1½ft (45cm) high, colouring up brilliantly in autumn. White, flat flowerheads are followed by generous crops of pink-flushed berries.

All these shrubs can be planted at any time of year from containers, but take care to ensure that they do not suffer from drought during their first summer.

CHAPTER 6

Alpines for Different Situations

The following lists are intended to offer you a bit of help and inspiration when searching for alpines to fill particular spots in your garden. Check out the information given about the plant in the A-Z section to make sure that it really fits the bill.

Plants for Shade

Adiantum, Ajuga, Andromeda, Anemone, Arisarum, Arum, Asarina, Asplenium, Blechnum, Cornus, Epimedium, Haberlea, Hacquetia, Hepatica, Lewisia, Lysimachia, Mertensia, Phyllodoce, Polygala, Pratia, Ramonda, Saxifraga, Shortia, Synthyris, Thalictrum, Trillium, Uvularia, Viola, Wulfenia.

Plants which Tolerate Chalky Soil

Acaena, Adonis, Aethionema, Alchemilla, Alyssum, Anthyllis, Arabis, Armeria, Arum, Asplenium, Astilbe, Aubrieta, Calandrinia, Campanula, Cerastium, Cichorium, Convolvulus, Daphne, Dianthus, Diascia, Dryas, Edraianthus, Erodium, Erysimum, Geranium, Globularia, Gypsophila, Hacquetia, Helianthemum, Hepatica, Hypericum, Iberis, Jasione, Lychnis, Mimulus, Oenothera, Onosma, Origanum, Penstemon, Phlox, Phyteuma, Potentilla, Pulsatilla, Saponaria, Saxifraga, Scabiosa, Sedum, Sempervivum, Thymus, Verbascum, Veronica, Zauschneria.

Plants to Grow in Walls

Arabis, Asarina, Asperula, Aubrieta, Calandrinia, Campanula, Cerastium, Cymbalaria, Erigeron, Euphorbia, Genista, Gypsophila, Helianthemum, Hieraceum, Iberis, Leucanthemum, Lewisia, Linaria, Lithospermum, Lychnis, Oenothera, Onosma, Othonnopsis, Papaver, Penstemon, Polygonum, Saponaria, Saxifraga, Scabiosa, Scutellaria, Sedum, Sempervivum, Tanacetum, Teucrium, Tropaeolum, Zauschneria.

Plants for Scree Beds

Acantholimon, Anacyclus, Anchusa, Androsace, Aquilegia, Asperula, Calandrinia, Callianthemum, Carlina, Celsia, Cichorium, Codonopsis, Convolvulus, Daphne, Douglasia, Dryas, Edraianthus, Erigeron,

Erinacea, Eriogonum, Erodium, Euryops, Frankenia, Geum, Globularia, Helichrysum, Hypericum, Leontopodium, Leptospermum, Leucanthemum, Linaria, Minuartia, Myosotis, Nierembergia, Ononis, Onosma, Origanum, Papaver, Raoulia, Sedum, Sempervivum, Silene, Teucrium, Verbascum, Zauschneria.

Plants for Peat Beds

Andromeda, Anemone, Arcterica, Arctostaphylos, Bruckenthalia, Cassiope, Cortusa, Corydalis, Cryptogramma, Cypripedium, Epimedium, Galax, Jeffersonia, Linnaea, Meconopsis, Mertensia, Ourisia, Phyllodoce, Polygala, Sanguinaria, Shortia, Synthyris, Trillium, Uvularia.

Plants for Troughs and Sinks

Androsace, Antennaria, Asperula, Calceolaria, Dianthus, Douglasia, Draba, Edraianthus, Erinus, Eriogonum, Helichrysum, Hutchinsia, Kelseya, Leontopodium, Mazus, Minuartia, Myosotis, Ononis, Phyteuma, Plantago, Polygala, Raoulia, Sagina, Saxifraga, Sempervivum, Soldanella, Thymus.

Plants for Patios and Paths

Acaeana, Antennaria, Anthemis, Arabis, Arenaria, Cotula, Cymbalaria, Dianthus, Erinus, Erysimum, Euphorbia, Frankenia, Globularia, Gypsophila, Helichrysum, Hieraceum, Iberis, Linaria, Lithospermum, Lychnis, Mentha, Minuartia, Oenothera, Papaver, Polygonum, Saxifraga, Sedum, Sempervivum.

Plants for Moist Soils

Ajuga, Anagallis, Arenaria, Astilbe, Cornus, Dodecatheon, Mimulus, Parochetus, Pratia, Primula, Ranunculus, Viola, Wulfenia.

Plants for Alpine Lawns

Acaena, Antennaria, Anthemis, Carlina, Coprosma, Cotula, Crepis, Cymbalaria, Dryas, Festuca, Frankenia, Globularia, Mazus, Mentha, Polygonum, Pratia, Raoulia, Sagina, Thymus.

Useful Addresses

Specialist Societies

Not only are the following societies useful sources of information, but they are also able to supply seeds that cannot be obtained elsewhere.

Alpine Garden Society, Lye End Link, St John's, Woking, Surrey
Northern Horticultural Society, Harlow Car Gardens, Harrogate, North Yorkshire
Royal Horticultural Society, Vincent Square, London SW1
Scottish Rock Garden Club, 21 Merchiston Park, Edinburgh EH10 4PW
American Rock Garden Society, Box 282, Mena, Arkansas, 71953, USA

Alpine Nurseries

Jim Archibald, Buckshaw Gardens, Holwell, Sherborne, Dorset
S. W. Bond, Thuya Alpine Nursery, Glebelands, Hartpury, Gloucester
Bressingham Gardens, Diss, Norfolk
Broadwell Alpines, Broadwell, Moreton-in-Marsh, Gloucestershire
Beth Chatto, Unusual Plants, White Barn House, Elmstead Market, Colchester, Essex
P. J. and J. W. Christian, Pentre Cottages, Minera, Wrexham, Clwyd, N. Wales
County Park Nursery, Essex Gardens, Hornchurch, Essex
Jack Drake, Inshriach Alpine Plant Nursery, Aviemore, Inverness-shire, Scotland
Edrom Nurseries, Coldingham, Eyemouth, Berwickshire, Scotland
Hartside Nursery Garden, Low Gill House, Alston, Cumbria
Holden Clough Nursery, Holden, Bolton-by-Bowland, Clitheroe, Lancashire
W. E. Th. Ingwersen, Birch Farm Nursery, Gravetye, East Grinstead, West Sussex
Reginald Kaye, Waithman Nurseries, Silverdale, Carnforth, Lancashire (Fern specialist)

Useful Addresses

Dwarf Bulb Suppliers

Avon Bulbs, Bathford, Bath
Broadleigh Gardens, Barr House, Bishop's Hull, Taunton, Somerset
de Jager, The Nurseries, Marden, Kent
Potterton and Martin, The Cottage Nursery, Moortown Road, Nettleton, Caistor, Lincolnshire

Bibliography

Alpine Garden Society Bulletins: 1933-82
Bartlett, M., *Gentians*. Blandford, 1975
Bawden, H. E., *Dwarf Shrubs*. Alpine Garden Society, 1980
Bloom, Adrian, *Conifers for Your Garden*. Floraprint, 1972
Brickell, C. D. and Mathew, B., *Daphne*. Alpine Garden Society, 1976
Chatto, Beth, *The Dry Garden*. Dent, 1978
Clay, Sampson, *The Present Day Rock Garden*. Nelson, 1954
Elliott, Clarence, *Rock Garden Plants*. Arnold, 1935
Elliott, Joe, *Alpines in Sinks and Troughs*. Alpine Garden Society, 1974
Elliott, Roy, *Lewisias*. Alpine Garden Society, 1974
Evans, Alfred, *The Peat Garden*. Dent, 1974
Farrer, Reginald, *The English Rock Garden*. Nelson, 1948
Farrer, Reginald, *My Rock Garden*. Edward Arnold, 1908
Green, Roy, *Asiatic Primulas*. Alpine Garden Society, 1976
Grey-Wilson, Christopher, *Dionysias*. Alpine Garden Society, 1969
Grey-Wilson, C. and Blamey, M., *The Alpine Flowers of Britain and Europe*. Collins, 1979
Griffith, Anna. N., *Collins Guide to Alpines*. Collins, 1964
Harding, Winton, *Saxifrages*. Alpine Garden Society, 1970
Heath, Royton, *Collector's Alpines*. Collingridge, 1964
Hellyer, Arthur, *Collingridge Encyclopedia of Gardening*. Hamlyn, 1976
Hills, L. D., *The Propagation of Alpines*. Faber & Faber, 1950
Ingwersen, Will, *Manual of Alpine Plants*. Ingwersen & Dunnsprint, 1978
Jackson, Robert, *Gardening on Chalk and Lime Soil*. William's & Norgate, 1941
Keble Martin, W., *The New Concise British Flora*. Michael Joseph & Ebury Press, 1982
Mansfield, T. C., *Alpines in Colour and Cultivation*. Collins, 1945
Mathew, Brian, *Dwarf Bulbs*. Batsford, 1973
Saunders, D. E., *Cyclamen*. Alpine Garden Society, 1976
Smith, G. F. and Lowe, D. B., *Androsaces*. Alpine Garden Society, 1977
Wilkie, David, *Gentians*. Country Life, 1950

General Index

Numbers in italic refer to line drawings.

Common Name Index